QoS in Integrated 3G Networks

For a listing of recent titles in the *Artech House Mobile Communications Series,* turn to the back of this book.

QoS in Integrated 3G Networks

Robert Lloyd-Evans

Artech House
Boston • London
www.artechhouse.com

Library of Congress Cataloging-in-Publication Data
Lloyd-Evans, Robert.
    QoS in integrated 3G networks / Robert Lloyd-Evans.
        p.   cm. — (Artech House mobile communications series)
    Includes bibliographical references and index.
    ISBN 1-58053-351-5 (alk. paper)
        1. Global system for mobile communications.   2. Wireless communication
    systems—Quality control.   I. Title: Quality of service in integrated 3G networks.
    II. Title.   III. Series.
    TK5103.483 .L56   2002
    384.5—dc21   2002021595

British Library Cataloguing in Publication Data
Lloyd-Evans, Robert
    QoS in integrated 3G networks. — (Artech House mobile
    communications series)
    1. Mobile communication systems
    I. Title
    621.3'8456

TK
5103
·483
L67

    ISBN 1-58053-351-5

Cover design by Igor Valdman

Further use, modification, or redistribution of figures, tables, and other materrial cited
in this book and attributed to ETSI (European Telecommunications Standards Institute)
is strictly prohibited. ETSI's standards are available from publications@etsi.fr, and
http://www.etsi.org/eds/eds.htm.

UMTS is a trademark of ETSI registered in Europe and for the benefit of ETSI members
and any user of ETSI Standards. We have been duly authorized by ETSI to use the word
UMTS, and reference to that word throughout this book should be understood as UMTS™.

cdmaOne is a trademark of the CDMA development group.

International Standard Book Number: 1-58053-351-5
Library of Congress Catalog Card Number: 2002021595

10 9 8 7 6 5 4 3 2 1

Contents

v

Preface

This book is intended to provide a self-contained general understanding of the factors that determine quality of service (QoS) in a third-generation (3G) mobile network that interacts with other fixed networks. Since QoS is an end-end quantity, and the mobile networks interact both with each other and with fixed networks, the discussion includes topics applicable to all types of networks.

The style of presentation here is aimed at network engineers for both mobile and fixed networks with an interest in QoS. Both categories of engineer are likely to require an overall understanding of these issues, since mobile engineers will have to investigate users' problems on accessing remote hosts and services, while engineers who support corporate networks will be expected to solve problems when staff members use their 3G phones to access those networks. The depth of treatment is intended to provide a general understanding of the complete range of topics, which should enable the reader to see what is important in any given situation, and to be able to use detailed specialized references on individual topics. The book is also suitable for students who need a general survey of these topics.

Many network engineers possess a very detailed knowledge of specific ranges of equipment but lack a comprehensive understanding of the principles involved. It is this gap that is addressed by this volume in relation to QoS. There are several thousand different recommendations and standards applicable to these integrated networks, together containing more than a million pages of material—engineers are lucky if they are given the time to read a thousand pages. With this in mind, this book provides sufficient references to enable engineers to home in rapidly, when necessary, on those documents that contain the details of quality-of-service issues.

Mathematics is avoided with the exception of Chapter 2 on coding, where a minimum is used to describe an essentially mathematical subject that is fundamental to the operation of 3G mobile networks. Elsewhere the emphasis is on data structures and protocols, as these are the features that a field engineer normally has to examine. Chapter 2 is important for junior design engineers but can be omitted or just skimmed for buzzwords by field engineers.

Chapters 3, 4, and 5 provide an overview of the radio technologies that are applicable to 3G networks. The emphasis is placed on data structures, timing, and signaling, as these are the factors most pertinent to QoS.

Chapter 6 provides an overview of radio network design, capacity, and planning. This chapter is primarily for the benefit of engineers from fixed network backgrounds who are likely to be unfamiliar with the propagation and design issues applicable to the radio access network.

Chapter 7 describes the network protocols that are used in both the radio network and the fixed networks with which it interacts. Emphasis is placed on those aspects of the protocols that are most vital to QoS, covering both signaling and traffic transfer. This chapter should also be useful to engineers responsible for quality issues on purely fixed networks. Material in this chapter is important background for the subsequent chapters.

Chapter 8 describes the interfaces between the radio access network and its associated core network, in addition to gateways to external networks and network management. This covers early ideas on the Internet multimedia core in addition to the circuit switch and packet switch network cores.

Chapter 9 deals with applications at a fairly general level, based on the UMTS classification of application types. It indicates the expectations of quality targeted by 3G mobile networks and how the required quality is signaled. Finally, Chapter 10 provides a more detailed look at the main applications for which QoS is most critical. These are voice and video features, and the chapter describes the compression algorithms used and the issues involved in their network transport.

While the book is aimed at engineers working with 3G networks in some form, Chapters 7, 9, and 10 provide a stand-alone guide to QoS that is also applicable to fixed networks in isolation.

The author thanks the Third Generation Partnership Project and the Third Generation Partnership Project Two for permission to use some material from their interim specifications in this book.

1

Introduction

1.1 Evolution of Mobile Networks

The first mobile networks were analog systems, mostly introduced in the 1980s. The specifications varied according to their geographic location: typical examples being Advanced Mobile Phone System (AMPS) in the United States, Total Access Communications System (TACS) in Europe, and Nordic Mobile Telephone (NMT) in Scandinavia (NMT was the very first in 1979). Collectively, these are now usually referred to as first-generation (1G) systems and were mainly used for voice, although data communications was also supported. The maximum bit rate for data was usually a nominal 2,400 bps. Due to high error rates, however, forward error correction (FEC) usually had to be employed, resulting in an effective rate for user data and its protocol overheads often as low as 1,200 bps. At these low rates very few data applications were practical, and the services were used for little more than paging, and for service engineers to download small data files. Voice quality was also poor because of areas of poor reception, network congestion, and the high degree of voice compression employed. This poor quality had an important side effect: it started a process of conditioning users to accept much lower voice standards than had been the norm on both PSTN and private corporate networks, so paving the way for the burgeoning Voice over IP (VoIP) services that are now appearing.

From the outset, the European Telecommunications Standards Institute (ETSI) envisaged the development of mobile telephony as a three-stage process, and the next major step in evolution was the appearance of digital mobile networks—the so-called second generation (2G). Once again the standards varied throughout the world, with global service mobile (GSM) being the norm in Europe and parts of Asia, while in the Americas and parts of Asia three further standards became widespread: North American/United States digital communication (NADC/USDC), Digital Advanced Mobile Phone System (DAMPS), and code division multiple access (CDMA). GSM and NADC are based on time division multiple access (TDMA) and offer better speech quality, more security, and global roaming than the earlier analog systems. Data in 2G systems is supported at up to 9.6 Kbps generally, and nominally at up to 19.2 Kbps (less in practice) by the Cellular Packet Data Protocol (CPDP), which taps unused speech cell capacity in DAMPS. In addition to the differences in technology between the services, there is a further administrative difference between the European and American systems. The two approaches use different ways of describing the subscriber's identity and use different protocols for transmitting such information: GSM-Mobility Application Part (GSM-MAP) in Europe, and American National Standards Institute standard 41 (ANSI-41) in America. As a result, gateways are required to handle administrative matters between the two types. Detailed descriptions of most of these 1G and 2G systems may be found in general textbooks [1].

GSM is by far the most widespread of the 2G technologies. Out of roughly 600 million handsets in 2001, about 70% were GSM, with approximately 10% each for American TDMA and CDMA systems. The most sophisticated of the 2G systems, however, is CDMA, wherein a single voice or data message is spread over multiple frequencies by means of a spreading code that is unique within a cell. Spread spectrum systems were originally developed for military use to provide immunity from jamming, and their introduction to mobile networks as a means of controlling interference was pioneered by Qualcomm. This system has several times the capacity of the others, partly owing to its use of silence suppression for voice and the ability to use the same frequency in adjacent cells and sectors, and it has been progressively developed under the American IS-95 standards. The earliest versions, IS-95A and IS-95B, are frequently referred to under the trademark cdmaOne. Handsets for each of these versions can operate on a network based on any of the others, but with limits on their performance. In each of these systems a user can only make a single call at a time, so the range of applications is still primarily confined to voice, with the most important data functions being file downloads by peripatetic personnel (e.g., traveling businessmen) and text messaging by teenagers and young adults.

The advent of the World Wide Web (WWW, or the Web) on the Internet has led to the development of services designed to provide easy access to it over these 2G phones—notable examples are Wireless Application Protocol (WAP) for GSM and I-Mode by NTT DoCoMo in Japan. I-Mode has proven extremely popular in Japan, where the low penetration of personal computers (PCs) within the population has made it the most widespread form of access to the Internet despite its low data rate of 9.6 Kbps, with a current fad for downloading cartoons. European WAP has been less successful and is being replaced by M-services, which has a better user interface. Another factor in the relative lack of success of WAP compared to I-Mode has been the failure of providers of WAP services to enter into joint projects with application developers.

In order to provide a more satisfactory Internet service, two main developments are required: (1) much higher data rates, and (2) uninterrupted Web access while making voice calls. The first of these is partially addressed by the so-called 2.5G technologies: high speed circuit-switch data (HSCSD), general packetized radio service (GPRS), and enhanced data rates for GSM evolution (EDGE), as well as cdmaOne. These offer Internet access at higher rates than the typical dial-up modem of a PSTN user and roughly comparable to basic-rate Integrated Services Digital Network (ISDN). GPRS and EDGE use the same TDMA technology as GSM at the physical level and are much cheaper for a GSM operator to deploy than going over to the totally new CDMA technology, wideband CDMA (WCDMA). GPRS phones can belong to one of three classes, the most basic of which is effectively WAP or M-service functionality at up to 76 Kbps, and the best of which allows a user to suspend a data transaction temporarily while taking a voice call. HSCSD is more basic than GPRS and is effectively a combination of several (often three) GSM interfaces in a single device in order to provide relatively fast downloads from WAP sites. EDGE uses a different modulation scheme to GSM that provides three times the bit rate, and is applicable to both enhanced circuit-switched data (ECSD) and enhanced GPRS (EGPRS). From a user perspective, the vital distinction between HSCSD and GPRS is that the former uses the same charging principles as GSM (i.e., charging per unit duration of the call) while GPRS is permanently on, but charged according to the volume of data. The always-on feature of GPRS also means that it gives slightly quicker log-on to a Web site than does HSCSD.

Ideally, the access rate should be higher still and the second criterion met, so the International Telecommunications Union (ITU) proposed the IMT-2000 scheme to achieve this worldwide by using a common frequency band that would enable a single handset to be used for access everywhere.

The European version of this proposed service is called Universal Mobile Telecommunications Service (UMTS). This aims to support multimedia (such as video-conferencing) and provide access to the Internet for both mobile and static users alike at practical speeds. Provision of adequate quality is harder for high speeds of both motion and data, so IMT-2000 recommends three different categories with maximum data rates as shown in Table 1.1.

The basic technology selected is a CDMA scheme (WCDMA) in the 1.9- to 2.1-GHz frequency band (see Figure 1.1), with licenses being auctioned for this purpose by governments in Europe and Asia. In the United States, part of this frequency band had already been licensed for personal communications systems (PCSs), so a different standard is also proposed allowing use of other mobile radio frequencies. This standard is cdma2000: it operates over multiple carriers with frequency bands of 450, 800, 900, 1,800, and 1,900 MHz originally licensed for earlier services and is actually a group of standards characterized by the number of carriers that can be used simultaneously. For cdma2000 1X, there is just one carrier, while for the next version, cdma2000 3X, there are three, with other versions to follow.

The ITU has recognized four approaches that meet the minimum IMT-2000 standards: American cdma2000, formalized by the CDMA Development Group (CDG) [2]; two by the original Third Generation Partnership Project (3GPP) [3], namely WCDMA-FDD and WCDMA-TDD; and UWC-136HS from the Universal Wireless Communications Consortium (UWCC) [4]. 3GPP is an ETSI partnership initiative, and when its draft specifications are approved, they are published as standards by ETSI [5]. The main version of WCDMA is a frequency division duplex (FDD) option, while the third standard is a time division version of WCDMA (loosely related to DECT) instead of the main frequency division option. All but UWC-136HS use direct sequence (DS) spreading technology (DS-CDMA), but the TDD option uses the same frequency band for both uplink and downlink (necessitating time division) while the other two use

Table 1.1
IMT-2000 Mobility Categories

Category	Physical Speed	Data Rate
Limited mobility	Up to 10 km/h	2 Mbps
Full mobility	Up to 120 km/h	384 Kbps
High mobility	More than 120 km/h	144 Kbps

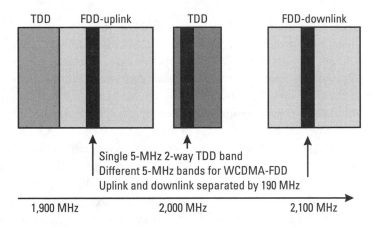

Figure 1.1 IMT-2000 frequency bands.

different frequencies for the two directions. The separation of channel frequencies is shown in Figure 1.1.

The bandwidth licensed for WCDMA-FDD is available throughout Europe and most of Asia, and that for WCDMA-TDD is available in most of Europe, while in America the WCDMA-FDD uplink bandwidth is already used.

The choice of which 3G technology for an operator to use is based on both technical and commercial decisions. Commercially there are significant advantages in going with the dominant standard, which is the GSM/UMTS path. Many users require roaming over other networks while away from home, and so need handsets that can operate on the majority standard. A second advantage is that application developers will concentrate on products that can interface to the dominant network type, although this is mitigated if a common network interface can be produced. Third, there should be economies of scale in production of handsets and base stations for the main standard.

On the other hand, upgrading a network from 2G to 3G is much easier in the case of cdmaOne/cdma2000 than for GSM/UMTS. As cdmaOne is a narrowband technology requiring 1.25 MHz per channel (whereas WCDMA uses 5-MHz channels from a new dedicated 200-MHz band), it is possible to transfer bandwidth gradually from a 2G service, such as D-AMPS, to the 3G services as demand for the latter increases rather than requiring the full bandwidth from the outset. This avoids the penalty of the high license fees paid by operators for the dedicated 3G bandwidth. The second point is that most of the basic technology remains the same when going from cdmaOne to full cdma2000, so that much of the initial network development consists of software upgrades to the base stations and controllers—although this advantage will be lost when extra base stations have to be added to provide increased capacity. The biggest distinction between cdmaOne and the first phase of cdma2000 is that the latter uses a more efficient form of modulation that provides twice as high a bit rate from the 1.25-MHz bandwidth. The transition from GSM to WCDMA via GPRS entails two major sets of upgrades. The first (and lesser) is an upgrade to GPRS where packet-switch (PS) technology is introduced but the physical radio interface is retained. In the second upgrade to WCDMA, the radio interface is totally changed. In addition, cdmaOne offers advantages over GPRS in the short term, notably through the use of Mobility IP as the core protocol, which allows users to contact any IP device. GPRS is unable to authenticate arbitrary IP addresses. WCDMA appears to offer significant technical advantages over cdma2000 in the long term, however, due to the wider bandwidth, better power control, and asynchronous operation.

The UWCC represents operators and vendors of TDMA equipment and has chosen to aim for interoperability primarily with the majority standard of GSM/UMTS. They support an interim body, the GSM ANSI-136 Interworking Team (GAIT), to promote interworking of GSM and IS-136 TDMA, and are developing UWC-136HS to give minimal 3G capability through the use of EDGE. This is likely to give its users the capability of about 100-Kbps throughput with bursts up to 384 Kbps, which should just be adequate for basic multimedia services. The UWC-136HS kit should be able to interwork with GSM/UMTS. The first step in the GAIT program is the production of sets that can support GSM at the frequencies available to the TDMA operators (such as 850 MHz). One commercial advantage of this approach is that these operators do not have to pay the high license fees that have been the norm for the UMTS frequency band.

Some operators may choose to move from their original standard to an alternative, and several such migration paths have been proposed. A few IS-95 CDMA operators in Japan and Korea have opted to move to UMTS.

The ability of 2.5G and 3G systems to provide adequate support for Internet applications depends on their ability to provide a suitable quality of service (QoS). The purpose of this book is to identify the requisite standards and to show the degree to which this is achieved and the means of doing so. The book concentrates on GSM/UMTS, but also deals at some length with cdma2000 and GPRS, and with only passing references to UWCC and the GSM/EDGE Radio Access Network (GERAN).

There are two main aspects of mobile services: mobility management and call control. The role of mobility management in determining call quality is comparatively limited, being restricted to registration of capabilities and subscriber rights together with handovers, so Chapters 3, 4, and 5 concentrate on call control, which is much more influential in this regard.

1.2 User Perception of Quality

The ultimate judge of network quality is the user, so it is essential to identify users' needs and then relate them to the technical standards of a network professional.

The first feature is access to the network. There are two main factors involved in this: The first is to have a very high chance of success, and the second is quick access. The first of these should be at least a 90% probability. The second is less critical and should be of the order of a few seconds. Users accustomed to dial-up access to the Internet may be accustomed to delays of half a minute, but a much shorter delay is highly desirable. Early 1G and 2G networks frequently failed to meet this access target due to inadequate resources for the numbers of users during periods of rapid growth.

The next essential criterion is that a call should remain up for the intended full duration. On a fixed network this is fairly easy to achieve, but in the case of a mobile network it is much harder as the call must be maintained during handovers and transit of areas of adverse topography and hence signal strength. As a fast-moving user may experience multiple handovers during a single call, the probability of failure on any one must be very low, perhaps in the region of 1%. In addition to the call staying up, there should be no significant glitches, loss of data, or disruption to speech during this process. These are some of the tougher aims to achieve on the network.

For calls involving speech, the next user criterion is that the voice quality should be good enough to be easily intelligible. In practice, this means that the compression algorithm should be adequate, that delay should not vary too much, and that the handover and reception issues above should be reasonable.

For data services, the most important single factor is usually the speed of access, followed by errors, loss of data, misdirection, and duplication. Whether errors are important depends on the type of service required; for example, loss or corruption of a few bits of data from a video stream will probably not be noticeable, but in an unsophisticated file transfer protocol (FTP), such as Trivial File Transfer Protocol (TFTP) or FTP, it could be very irritating. An even worse scenario for these simple protocols would be call termination before transmission was complete, as that would require the whole transaction to be restarted from scratch. A more sophisticated file transfer application, such as NFS, would have synchronization points to minimize this issue. This means that for a full understanding of these effects it is necessary to go into details, both of the specific application requirements and of the protocol capabilities. Thus, in the study of QoS, the IMT-2000 program defines four overall groups of UMTS applications: conversational, streaming, interactive, and background. These form the basis of the discussion of applications in this book (an additional chapter is also devoted to the influence of network protocols).

Many applications consist of several components with different needs. The simplest example of this is the downloading of a film, where it is necessary to synchronize the sound and video tracks, while the sound and video may each consist of several channels with distinct QoS requirements. In order to understand these effects, it is necessary to study the details of the application, so the final chapter is devoted to a study of some of the more important applications in greater depth.

1.3   Costs and Benefits of Quality

Quality is not just something that is nice to have; it has clear-cut cost and benefit implications. The simplest case to consider is the question of speech quality.

Suppose a CEO is holding a mobile phone conversation that has poor reception. From time to time a sentence will be unintelligible, forcing the CEO to request its repetition. Let's say that each such request and repetition will probably take about 10 seconds. If the remuneration package of the CEO is worth $10 million per annum for a 60-hour week, 50 weeks a year, then the rough cost of one of these incidents will be about $10 (twice that for two such CEOs talking to each other). The cost of the errors in a conversation lasting a quarter of an hour could be equivalent to the daily salary of a junior employee. Of course, not all allegations of poor quality of this type will be justified, as some CEOs will demand a repetition on the grounds of bad reception in order to gain more thinking time for a considered response—as such, network managers need objective tools for measuring voice quality.

Another example is the case of a sales clerk taking an order over the phone. If reception is poor, an error may be made that results in an incorrect set of goods being delivered. In this case, the costs of the error will be those for reprocessing the order and for the additional shipping charges incurred in returning the incorrect goods and delivering the correct set. As only minor orders are likely to be taken by phone, the cost of such an error will probably be of the order of $50, although bulky goods or a poor order-processing system would boost this. There is also the relatively unquantifiable cost of the loss of goodwill resulting from poor voice quality.

There is also a cost benefit to the operator of a CDMA network over that of a 2G TDMA network, and thus potentially to the customer, through the ability to support variable rates and save bandwidth during silences. The 3G variable rate codecs enable the quality to be improved by increasing the degree of error protection during periods of poor reception.

The benefit of good quality is most apparent in the case of medical image transfer. If an off-site top consultant has to be contacted urgently to interpret a brain scan or X ray after an emergency, then the ability to transmit a high-quality image can be a matter of life or death. To a loved one, no amount of money could compare to this benefit, but to an accountant working for an insurance company, the associated cash value could easily be a few million dollars. Conversely, a poor quality image might contain false information and lead to an incorrect diagnosis.

At a more mundane level where an employee is using a mobile handset to obtain corporate information from a Web site, rapid data transfer will provide a significant saving in labor costs. Suppose that the application requires the employee to think about each segment of data received and, furthermore, that the segment transfer time is not long enough to allow another task to be performed in the meantime. Reception of a 2-MB file at a throughput of 1 Mbps will take 16 seconds, whereas at 50 Kbps it will take 320 seconds—the higher speed will save about 5 minutes. This cost saving will quickly become significant if the employee performs many such subtasks during the day. This is an important issue in deciding whether GPRS will be adequate, or whether a full 3G service is needed, subject obviously also to the relative costs of the two services.

Even when the mobile is being used purely for entertainment, quality still has identifiable costs. If the handset is used for viewing a video on demand or any Internet pay-per-view option, then the cost of the service is wasted if the reception is so poor that the video is not enjoyable. A participant in an interactive game will be at a serious handicap if quality is poor.

The biggest benefit of adequate quality in the 3G network is simply that of mobility itself, so that an employee can tackle a wide range of tasks without having to be in the office. The additional benefits of 3G services over those of 2G in this respect are primarily those associated with high data rates for file transfer, rather than video applications. The most dramatic gains for mobile video occur for military applications where images can be sent or received via mobile base stations and satellite networks to gain a competitive edge over opponents. Once again, this is a life or death issue. Mobile video also assists in general security operations, enabling field personnel to receive images from security devices and thereby reach the scene of an incident faster.

A network operator also suffers if quality is poor, as that will lead to a loss of customers to rivals, although some of the loss in revenue may be mitigated if the sneaky practice of charging multiple retransmissions of errored packets as separate individual packets is used. Commercial considerations mean that an operator has to look for the best trade-off between quality and cost of service provision rather than pure service quality. In the case of mobile networks, the best quality for an individual user would result from a high signal-to-noise ratio (SNR), but this causes interference to other users, thereby reducing capacity and revenue; as a result, strict power control is used to ensure quality that is adequate for the application, but not unnecessarily high. The largest quality factor is simply the distinction between 2G and 3G services, and the high cost of providing the latter means that many operators will be forced to restrict its geographic coverage to areas of high user density.

In integrated 3G networks the mobile part is not the only section to influence quality. The next section provides an overview as to which factors apply in which domain of the network.

1.4 Influence on Quality of Different Parts of the Network

Quality is an end-to-end characteristic of a call to which each network constituent contributes. The parts of the network to consider are the radio link from the user equipment to the base transceiver station for the cell, the terrestrial radio network linking the cells and their controllers [the UMTS Radio Access Network (UTRAN) in 3GPP terminology], gateways to the core network, the core itself, and remote peripheral networks (fixed or mobile).

In addition, quality is also influenced by the ability of the overall network to signal individual call requirements in the control plane and to support these needs throughout the call in the user plane.

Signaling ability is needed at each stage in network transit. If the user equipment (UE) is unable to signal the quality needed for individual messages, then the best that can be done is for the mobile service provider to offer a subscription-based standard quality. The UTRAN then has the capacity to pass on the standard or signaled UE needs via the signaling ability of IP or ATM, as used, to the core network with conversion at the gateways. Finally, quality control to the called party will only be maintained over the last section if it too can signal quality needs. Failure of any part of the network to support appropriate signaling will mean some loss of quality in the user plane.

All sections of the overall network contribute to grade of service for initial access, but in the case of 2G networks the radio sections have given a lower probability of successful access than any fixed networks to which they have been attached. This has been the result of poor coverage in regions of unfavorable topography combined with shortages of resources. These factors will probably persist in 3G networks, and congestion of Web hosts may also be significant. Fixed networks are designed to have an extremely low rate of premature call termination, and handover difficulties in the mobile section are the main cause of this problem.

The initial radio link is the most critical section of the user plane in determining quality, but by no means is it the only significant part. The most extreme effects of this are high error rates compared to the rest of the network and potentially slow transmission as the rate may be from 9.6 Kbps up to 2 Mbps. The high error rates entail a need for transmission of data in blocks with some means of error correction, while the slow transmission rate contributes to long delays. The UTRAN possesses links of much higher speed and quality, but these can be associated with loss of data or even of calls at handover between cells, while congestion on these shared links will lead to variable queuing delays and resultant loss of quality due to jitter. Premature loss of calls is most probable when handovers are made that entail a change in carrier frequency. The core network is likely to consist of very high capacity fiber optics; unless they are overloaded they should only have one adverse effect, namely propagation delay on long links. This delay is typically about 1 ms per 100 miles: Delay on transcontinental or intercontinental terrestrial links is on the order of 30 to 50 ms but rises to 270 to 300 ms for hops via geostationary satellites.

The destination network is potentially a major source of quality problems. These can either be a mirror image of the above for another mobile network or else a different set characteristic of a fixed network. If the call destination is a Web site, then it is probable that it will be accessed by links of high capacity and quality but with the likelihood of congestion and variable delay. If the destination is an IP phone or individual data user, then low speed data links may be involved, leading to considerable and variable delays.

VoIP, in particular, has its own quality problems, and these factors are considered in a later chapter. For circuit-switch (CS) connections, there can be a considerable loss of quality due to transcoding when the voice compression algorithm is different at the two ends of the call. This problem is minimized by having a default codec to use in the 3G networks, plus options for attempting codec negotiation otherwise, but remote plain old telephone service (POTS) phones cannot participate in this and always require transcoding.

Loss of data is liable to occur on any part of the overall network that uses any form of statistical multiplexing of traffic from multiple sources. Asynchronous transfer mode (ATM) and frame-relay networks explicitly accept this loss and state rules for dropping traffic that exceeds an agreed subscriber rate, using discard eligibility or loss priority as a control mechanism. Likewise, routing switches on PS networks drop packets according to proprietary rules when they become congested. Traffic also makes use of a mixture of shared and dedicated channels over the radio links, with the former used for data of an intermittent bursty nature leading to possible losses during temporary overloads. Again, traffic priority rules can be applied that minimize the resultant damage.

Misdirection can occur anywhere that the address information is corrupted. This is most probable in the areas of high error rates, and hence on the radio link unless mechanisms are used to counter the problem.

Duplication of data is used by some protocols to prevent loss or corruption of data where delay requirements preclude the use of retransmission. Specific procedures have to be included to eliminate the duplicates, and this is frequently performed on the radio link.

In addition to the influence of the network in normal operation, QoS is also affected by security factors. The most direct source of loss of quality arises from denial of service attacks on a Web host to which the user is attached. The nature of these attacks depends on the operating system of the host, while the effects are congestion of either the network or of the host. While many of the defense mechanisms are dependent on the host, some, such as encryption, depend on the capabilities of the mobile also. Similarly, data integrity and authenticity are vital QoS requirements that are facilitated by support for encryption. Operational integrity of the mobile network also requires protection from viruses and from hacking into network control devices. One side effect of using encryption or ciphering is an increase in call setup time, typically of a few seconds, due to key exchange and synchronization.

1.5 Outline of the Book

The various aspects of QoS are described in the following chapters. Chapter 2 introduces the types of codes that are used on the radio links to provide error correction and to distinguish between different users and channels. Chapters 3, 4, and 5 describe the radio links for WCDMA, cdma2000, and GSM/GPRS, respectively. They do not attempt to give a complete coverage of the topics, but concentrate on those features that are directly related to QoS while largely ignoring other features such as mobility management. Chapter 6 contains a simple introduction to radio propagation and the sources of errors and shows how capacity is determined as well as the influence of QoS on the number of calls that can be supported and on cell size. QoS is an end-to-end concept and depends on how well it can be supported in the fixed core of the mobile network and upon any remote networks involved in a call. The level achievable depends on the network protocols that are used in these regions. Chapter 7 lists the main features of the standards and recommendations that determine this. Chapter 8 then describes the interfaces between the radio access network and both the core and remote networks, again with special emphasis on QoS. In Chapter 9 attention is turned to the classes of applications that are used on 3G networks, the intended QoS targets, and the contributions from different parts of the overall network. The final chapter describes how the main voice and video applications work.

References

[1]   Miceli, A., *Wireless Transmission Handbook*, Norwood, MA: Artech House, 2000.

[2]   CDMA Development Group, http://www.cdg.org.

[3]   Third Generation Partnership Project, http://www.3gpp.org.

[4]   Universal Wireless Communications Consortium, http://www.uwcc.org.

[5]   ETSI, http://www.etsi.org.

2

Coding Overview

2.1 Generalities

From the earliest days of the information theory pioneered by Shannon [1], it was realized that transmission of a continuous bit stream was not the best way to send a message; instead, more robust symbols or blocks of data should be used [2, 3]. Prior to Shannon's work the main approach to improving signal quality was by increasing the SNR, but Shannon's formula

$$\text{Theoretical capacity} = \text{bandwidth} \times \log_2 (1 + \text{SNR}) \qquad (2.1)$$

showed that the error rate could be made as small as desired for a given SNR by increasing the bandwidth, although all practical systems fall short of this limit. The ability to increase quality without increasing the signal strength is particularly important in mobile phone networks as it reduces the potential interference between users.

These principles were recognized in the development of character- and bit-oriented protocols [e.g., binary synchronous communications (BSC), synchronous data link control (SDLC), and high-level data link control (HDLC)] with error check fields to allow identification and partial correction of blocks containing errors so that these could be retransmitted. Mobile radio links are subject to even higher error rates than early analog lines, with the result that error correction by retransmission is no longer always adequate, making FEC also necessary. FEC is also essential where information is broadcast, as retransmission is then totally impractical. This is achieved through the conversion of the initial bit stream into transmission codes having this FEC property. CDMA networks also use spreading codes to distinguish between different cells, users, and channels and spread the signal over the range of frequencies in use. Use of transmission and spreading codes is fundamental to CDMA networks.

In addition to transmission and spreading codes there are also two other types of code used in mobile networks: source codes and security codes. Source codes have a big impact on network performance because they provide data compression—examples of these include the Moving Pictures Expert Group (MPEG) standards for video compression and numerous voice compression standards such as G.723 and G.729. Security codes are the encryption codes used to ensure data confidentiality and have little influence on quality or performance, so are only discussed in outline in this book.

The emphasis in this chapter is on a rudimentary description of transmission and spreading codes, while source codes for voice and video are discussed in later chapters on applications.

2.2 Block Codes

Transmission codes can be grouped into several generic types, such as block codes and convolution codes. In addition to these generic distinctions, transmission codes in mobile phone networks are also classified according to their function (e.g., spreading codes and channel codes). In this chapter, transmission codes will first be outlined generically, then in later paragraphs by their function in 3G networks.

The most elementary transmission codes are the block codes. A block code C defines a unique map of a block consisting of k information symbols $\{i_0, i_1, \ldots, i_{k-1}\}$ into a code word of length n consisting of the n symbols $\{c_0, c_1, \ldots, c_{n-1}\}$. The fact that $(n - k)$ extra redundancy symbols are now transmitted in the coded block provides the possibility both for detection of errors and the correction of many of them. The downside of this is, of course, an increase in bandwidth, which is characterized by the code rate that is defined to be the ratio k/n, indicating the raw bandwidth as a fraction of the coded width. In general, this is beneficial whenever more than about 10% of the raw blocks would otherwise contain errors, but this has additional, specific advantages in CDMA in relation to the use of similar frequency bands in neighboring radio cells.

In order to provide freedom from errors, it should be as easy as possible for the remote decoder to distinguish between the various coded symbols that it receives. There is a quantitative measure of this, the Hamming distance, named after its founder [2]. The distance between two code words, **a** and **b**, belonging to the code C is the number of symbols by which they differ (i.e., a number in the range 0 to n). The minimum value of this distance for all pairs of code words in C is the Hamming distance, d. A block code C with these characteristics is usually denoted by the form $C(n,k,d)$ to illustrate its features.

For the simplest group of block codes (i.e., linear block codes, where linear means that adding two code words together automatically gives a third code word for the code), the code $C(n,k,d)$ is able to detect up to $(d-1)$ errors and to automatically correct up to $e = [(d-1)/2]$ of them. By way of illustration, a completely unencoded message would have $d = 1$ and so not be able to detect or correct any errors at all. The simplest examples in data communications of the use of block codes are the parity checks and the cyclic redundancy check (CRC) at the end of HDLC frames. The additional parity bit attached to an unencoded block boosts the distance from 1 up to 2, so that one error can be detected, but none is corrected, whereas the larger CRC fields allow some correction in addition to detection. The code words of $C(n,k,d)$ consist of k information bits and $(n-k)$ redundancy bits.

The number of possible code words in $C(n,k,d)$ is q^k, where q is the number of possible values for the raw input symbols and k is the number of symbols in the unencoded blocks as before [mathematically, q is the order of an underlying Galois field GF, (q), but this is not important to a rudimentary understanding of codes].

One of the simplest codes applicable in data communications is the 1-bit parity code, $C(k+1,k,2)$. Each information block consists of k bits, while the output consists of the original information bits together with an additional parity bit, c_k that is defined by the relation

$$\sum_{j=0}^{k-1} u_j + c_k = 0 \bmod(2) \tag{2.2}$$

In order for a code to be practical in data communications, it needs to be quick and easy to decode. The simplest such mode is based on the use of trellises.

A lot of sophisticated mathematics goes into the development of practical codes. Two of the most important general types used in data communications are the Bose-Chaudhuri-Hocquenghem (BCH) codes that form the basis of CRC error protection, and the Reed-Solomon (RS) codes that are used in more severe conditions. The BCH codes are designed to protect single bits, while RS codes aim to protect groups of bits and so handle burst errors.

In general a code can be expressed in several ways, but the most common in data communications is the so-called systematic form, where the output code word is expressed as the original input word followed by the error-correcting redundancy bits. The addition of the CRC field at the end of a data frame is a typical example of using the systematic form of a BCH code.

2.3 Trellis Codes

Trellis codes, which include the block codes, can be represented graphically by a trellis. A trellis $T(V,E)$ of length n consists of a set of vertices, V, and a set of directed edges, E. The vertices V can be split into $n + 1$ levels, t, $0 \leq t \leq n$, where n is the length of the code $C(n,k,d)$. The first and last vertices (i.e., those with $t = 0$ and $t = $ n) are special and are called the root and goal, respectively. An edge is a connection from a vertex of level t to one of level $(t + 1)$. A path is a sequence of edges that connects one vertex to another. Each edge is labeled by a symbol $c(e)$ from the set of q values that identifies the next symbol in the input word or redundancy components. A path defined by a sequence of m edges thus corresponds to a vector given by the components $c(e_1)$ to $c(e_m)$. A code trellis for the code $C(n,k,d)$ then has the following properties: Each path from the root to the goal has a corresponding vector that belongs to the code C, and to each code word **c** there corresponds at least one path from the root to the goal. Figure 2.1 illustrates

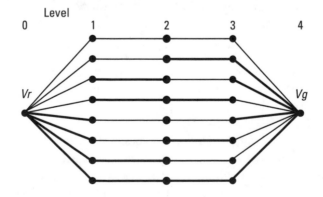

Figure 2.1 Trivial trellis for $C(4,3,2)$.

one of the simplest possible cases, the parity check code $C(4,3,2)$, which has eight paths corresponding to the eight valid code words. This code has $q = 2$, corresponding to the two values 0 and 1, length 4 obtained by adding the parity status to the three input bits, and distance 2.

In this diagram, edges that correspond to an additional 0 input bit are shown as thin lines, while those corresponding to 1 are thick lines to represent labeling by these values.

There are multiple possible trellises for any given code in general, and that in Figure 2.1 is an example of a trivial trellis, where there is one path for each code word. The purpose of using the trellis is to reduce the computational complexity of decoding to the minimum possible, and this is achieved by minimizing the number of vertices in the trellis [4, 5]. Following the set of rules listed below produces a minimal trellis:

1. Starting from the left-hand side of the trellis (i.e., the root) with $t = 1$, merge all edges from vertices at level $(t - 1)$ that have the same level;

2. Repeat for the next higher level until the goal is reached at the right-hand side ($t = n$);

3. Working backwards from the goal with $t = (n - 1)$, merge all edges from vertices at level $(t + 1)$ that have the same label;

4. Repeat for the next lower level until the root is reached.

In the case of the parity check code $C(4,3,2)$, this results in the minimal trellis shown in Figure 2.2, again with thin and thick edges for 0 and 1.

The minimal trellis can be used to compare a received code word with all the possible values much more quickly than by a direct 1:1 comparison with all complete words.

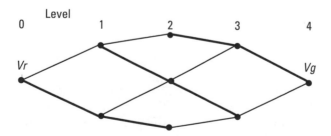

Figure 2.2 Minimal trellis for $C(4,3,2)$.

2.4 Viterbi Algorithm

The usual decoding scheme based on the trellis of a code is the Viterbi algorithm. This is an iterative process based on initialization and three other steps, as described below. Mathematically, it is similar to the Djykstra algorithm used for finding the shortest path in data communication routing protocols.

Initialization

> Put $t = 1$
>
> Set $B(v_R) = 0$, where v_R denotes the root vertex.

Step 1

For all edges **e** contained in E_t (the edges of level t connecting vertices of level $(t-1)$ to those of level t)

Set $B(f(\mathbf{e})) = B(I(\mathbf{e})) + L(\mathbf{e})$, where $L(\mathbf{e})$ is the branch metric for the distance of the tth element $c(\mathbf{e})$ of the code word from the corresponding element of the received word **r** and $I(\mathbf{e})$ and $t(\mathbf{e})$ are the initial and final vertices of the edge. Typically in binary codes this is the number of coordinates in **c** that agree with those of **r** up to this level.

Step 2

If more than one edge terminates at the vertex v' belonging to the set V_t, then select the path with the maximum value of $B(f(\mathbf{e}))$ as the survivor and assign the metric to the vertex $f(\mathbf{e})$. If all metrics are equal, then one is chosen at random.

> Put $t = t + 1$
>
> If $t = n$, then return to step 1,
>
> Else go to step 3.

Step 3

The decoded sequence **x** is given by the survivor paths that terminate at the goal vertex v_G.

This technique aims to provide the most likely decode of the received word **r**, but because of the assumed transmission errors, it cannot be guaranteed to be correct. The maximum number of coordinates that coincide corresponds to the minimum Hamming distance between **r** and any code word **c**. Use of the survivors only in the computation greatly reduces the amount of decoding work required for long codes. For example, if the received code word for $C(4,3,2)$ had been $(0,1,1,0)$ there would be no point in further comparison with codes that started $(1,0,\ldots)$. In the absence of errors, the Hamming distance between **r** and the selected **c** would be zero in this case, but it would not be possible to say what the correct decode was if any error was present.

There are several variations of the Viterbi algorithm to deal with more complicated cases, such as situations where the probabilities of receiving the various code words are unequal or for nonbinary codes [4, 5]. In general these variations use the same overall procedure, but with different types of branch metric.

The Viterbi algorithm works well for short codes, but eventually becomes impractical as *n* and the consequent trellis size increase. In mobile networks this is one factor that tends to limit the sizes of blocks transmitted. Section 2.5 contains an example (see Figure 2.3) of the use of the Viterbi algorithm for a very simple convolutional code.

2.5 Convolutional Codes

Convolutional codes were introduced to code continuous binary bit streams instead of finite discrete blocks (but they are also used for the latter). The

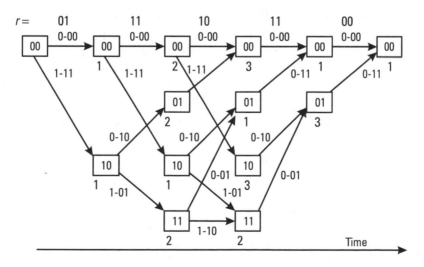

Figure 2.3 Coding/decoding.

fundamental difference between convolutional and block codes is that the latter have no memory, whereas the former do since the encoded output is determined not just by the current input block, but also by one or more of its predecessors. A related mathematical distinction is that the output of the block codes is related to the input by an algebraic formula, whereas there is no such simple relation for a convolutional code. Convolutional codes tend to be better at eliminating isolated random errors than the block codes, but their biggest advantage in mobile networks is the ease by which they can be modified to provide matching of the output data rate with physical channel requirements (see Section 2.6).

A continuous binary (i.e., in terms of the block code jargon above, $q = 2$) input information stream $\mathbf{u}$ is partitioned into blocks $\mathbf{u}_a$ consisting of k bits each. Each such block is then converted by the encoder to an output block $\mathbf{v}_b$ consisting of n bits, much as for the block codes, except that this output depends on the previous m blocks as well as the current input,

$$\mathbf{v}_t = F\left(\mathbf{u}_{t-m}, \mathbf{u}_{t-m+1}, \ldots, \ldots \mathbf{u}_t\right) \tag{2.3}$$

where t indicates the sequence number.

The corresponding convolution code $C(n,k,|m|)$ is said to have memory m and rate k/n (as for a block code) and consists of all the sequences that can be produced by an encoder of this type.

In effect there is one generator for each of the n bits that make up a code word. These generators give the code its name (i.e., convolutional) because they are represented mathematically as the convolution product of the input bits with the responses of the encoder to single bits subject to delays of up to $(m-1)$ bit times. This can be expressed formally as

$$\mathbf{v}^j = \Sigma_{i=0}^{k-1} \mathbf{u}^i * g_i^j \tag{2.4}$$

where g is the generator matrix for the code and the summation runs over the elements of the input block. The significance of the term *convolution* is shown in the explicit form of (2.4), showing the time components with the last input bit being combined with the earliest generator component

$$\mathbf{v}_t^j = \Sigma_{i=0}^{i=k-1}\left(\mathbf{u}_t^j \times g_{i,0}^j + \mathbf{u}_{t-1}^j \times g_{i,1}^j + \mathbf{u}_{t-2}^j \times g_{i,2}^j + \ldots + \mathbf{u}_{t-m}^j \times g_{i,m}^j\right) \tag{2.5}$$

Some of the g coefficients will, in general, be zero, and the value of m may vary according to which output bit j is being considered. The individual values of *m* for the separate output bits are called the constraint length for the generator of that bit. The memory is then defined to be the largest of the different values, so that the last few *g* coefficients will be zero for bits with constraint length less than this maximum.

Equation (2.5) can also be transformed into a form determined by a delay operator, *D*, instead of the time, where the power of *D* represents the bit delay relative to the initial bit.

$$V^{j}(D) = \Sigma_{i=0}^{k-1} U_{i} \times G^{ij}(D) \qquad (2.6)$$

The elements of the matrix **G** are then sums of powers of the delay operator *D*, and this form of expression is very widely used. A further alternative is to describe the elements of the matrix via either a binary or octal representation of the powers of *D*.

The sum of the individual constraint lengths for the output bits is called the total constraint length, and sometimes also confusingly referred to as the memory of the encoder since it is equal to the number of memory storage elements of the encoder. If this total constraint length is denoted by *M*, then the encoder has 2^{M} possible memory states. On receiving a new input block, the encoder goes from one memory state to another in addition to generating the output dependent on the new input block and the current memory state.

The operation of the encoder can be described via a state diagram for these memory states. This is a graph whose nodes represent the possible memory states: At time *t*, the encoder is in state S_{t}, when it receives an input block $\mathbf{u}_{t}$, which it converts in state-dependent manner into an output $\mathbf{v}_{t}$, and moves into a new state S_{t+1} resulting from the inclusion of $\mathbf{u}_{t}$ in its memory and the dropping of the earliest stored block. The transitions from one node to another are then labeled by the pair of blocks $\mathbf{u}_{t}$, $\mathbf{v}_{t}$.

The code words that can be generated by the code can also be represented diagrammatically by a code tree. This tree has a number of nodes that grows exponentially with time, making it impractical to use, but it does contain a number of periodically repeating subtrees. These are based on the fact that each node in the code tree can be associated with a memory state S_{t} that is uniquely defined by the path to that node. The evolution of the code depends only on the memory state and subsequent inputs, so identical paths branch out from similar memory states. Once *t* exceeds the memory parameter *m* (i.e., after the first few blocks), then transitions from nodes with equal states can be merged. The significance of this is that it produces a trellis representation of manageable size that can be decoded easily by the Viterbi algorithm, as in the case of the block codes.

The matrix $\mathbf{G}(I)$ defines a simple example of a convolutional code of rate 1/2 and memory 2:

$$\mathbf{G}(D) = (1 + D + D^2, 1 + D^2) \qquad (2.7)$$

That can also be expressed as $\mathbf{G}_0 = 7$ (octal) = 111 (binary), $\mathbf{G}_1 = 5$ (octal) = 101 (binary). The manner in which inputs are coded and decoded using this code is illustrated in Figure 2.3 for the case of a binary stream broken into blocks of three input bits (either zero or one) terminated by two zero tail bits to reinitialize the encoder to its starting state at the end of each block. At any time the encoder stores two input bits, denoted [AB] in the diagram, that defines its states. As each bit is either zero or one, there are four possible states: [00], [01], [10], and [11]. On receiving a additional input bit C, the encoder moves to the new state [CA] and outputs two bits X and Y; this input and output is denoted C-XY in the diagram, which shows all possible inputs and outputs for a single block with total output consisting of 10 bits per block. The diagram also shows how the decoder interprets an example of a coded sequence chosen as 0111101000 and shown as successive bit-pairs in bold on the top line of the figure. Starting at the left-hand side from the state [00], a metric is created showing the distance of each path from this coded sequence as successive bit-pairs are interpreted. The branch metric is the difference between the theoretical output bits and the received coded bit-pair, while the cumulative discrepancies by the best path to each state at any time is shown in bold below that state. The original input bits on the path with the lowest number of discrepancies define the decode of the received coded sequence. In the example, the first received bit-pair evidently contains an error, and the decoded sequence is 01000, of which the last two bits are just the standard tail bits.

2.6 Code Extension and Shortening

In communications, it is sometimes necessary to change the length of the code slightly in order to comply with some system constraint—such as the rate of a communications channel—and this is usually performed by repetition or deletion of symbols. Shortening is particularly important in this respect, and there are two ways of doing so:

1. Instead of the input symbol vector **u** consisting of the usual number k of symbols, only the first $(k - q)$ are selected, thereby reducing both the code length and dimension by the quantity q. This gives a new rate of $(k - q)/(n - q)$, but the minimum distance of the code is ether the same or greater than the original.

2. A number q of coordinates of each code word **c** are punctured (the technical term for deleted in this context) leading to a code rate of $k/(n - q)$. In this case the minimum distance is likely to be reduced, making it less effective for error correction.

Each of these methods leads to a new linear code, and the second method is extensively used with convolutional codes in CDMA to achieve data rate matching with line capacity. The degree of puncturing is sometimes changed within a radio frame according to the importance of the section of data being protected, so that vital data gets a lower error rate than less important data.

In principle there is a large number of possible punctured codes that can be derived from a given mother code so as to provide a given output rate, and a considerable amount of computer analysis is required to find one that gives the maximum distance between code words, and hence the best error correction. Puncturing is characterized by a deletion matrix that specifies which bits are dropped from the output stream.

2.7 Concatenation Codes

In order to increase the ability of a code to detect and correct errors, the distance between the code words needs to be increased. The easiest way to do this is simply to increase the length, n, of the code, but this makes decoding much harder. A better way to achieve this objective is to combine two or more codes to form a concatenation code. Concatenation codes belong to either of two categories: serial or parallel.

Serial concatenation codes are the more straightforward of the two types. They are generated by treating the output of one encoder as the input for the next of the two or more separate codes, so that the rate, R, of the concatenation code is the product of the rates, R_1 and R_2, of its components.

Each component code of a parallel concatenation code independently encodes the original data, or a permutation thereof, so that two or more separate code words are produced, each containing the same information bits but having different redundancy bits. The output of the parallel concatenation code consists of a single copy of the information bits together with the redundancy bits for each of the component codes. The rate R is related to its component rates by

$$1/R = 1/R_1 + 1/R_2 - 1$$

Very long codes can be constructed by concatenation of several short codes, but the complexity of decoding is just that of sequentially decoding the short constituent codes rather than the tougher problem of a single long code for which the Viterbi algorithm would be impractical.

A very important class of parallel-concatenated codes are the so-called turbo codes [6]. These use two convolutional codes in systematic form (often two copies of the same code) to produce a long code that can be easily decoded. These codes are also extremely efficient and allow a large reduction in SNR for a given quality as compared to an unencoded signal. They are widely used in CDMA for high-speed data transmission where small data blocks are impractical.

By combining two codes that are optimized for correcting different types of errors, one can obtain the best of both worlds with regard to quality. For example, block codes tend to be good at correcting burst errors (such as those resulting from a power surge or lightning strike) while convolution codes are better at handling isolated errors, so concatenating the two gives a code that is good at both.

Straightforward block codes or convolution codes treat each input symbol as being of equal importance, but this is not always the case. A very important group of exceptions are the instances when the input to the encoder is already the output of a source coder, such as voice or video compression. Most modern voice or video compressors produce output streams whose symbols have specialized significance, some being of critical importance while others are little more than filler. Dedicated concatenation codes can be constructed to match specific compression algorithms so that more error protection is provided to the symbols of critical importance than to those of minor importance, thus enabling high-quality transmission to be achieved without the huge overhead associated with giving high error protection to the entire data stream. In many instances the differing degrees of protection are produced by puncturing of the convolutional codes mentioned above.

Even with concatenation codes, burst errors can remain a problem. Another technique that is used to overcome this is the use of interleaving. Instead of transmitting the encoded blocks sequentially, they can be split into sections that are then interleaved so that sections of the first block are not transmitted until after some sections of the later blocks. This means that the effects of transient interference, such as Rayleigh fading or a lightning strike, are smeared over several blocks where they may be correctable rather than making one block unintelligible. Mathematically, the effect of interleaving is to increase the code distance. The downside to this is that all the interleaved sections have to be received and reordered before they can be decoded, thereby causing a significant delay. As convolutional coders can themselves sometimes produce burst errors, interleaving is also used between the different stages of a concatenated code.

2.8 Source-Coding Principles

The error correcting channel codes described above are all based on formal mathematics. The source codes used to provide traffic compression, however, are all based on much more informal methods. One of the most common methods used is Huffman or entropy coding, whereby the most frequently occurring discrete symbols (or sequences of symbols) are replaced by the shortest codes in a preset manner. This was widely used in data compression standards for early modems. It works best where the data stream is very repetitive, as is the case with control data for video frames for which it is used in algorithms such as the MPEG family (see Section 10.2).

Compression of long streams of nonrepetitive material, such as general text, is handled more efficiently by dictionary methods. A dictionary is compiled by the compressor, either as a preset table in the simplest cases, or more generally constructed dynamically during the course of transmission, and communicated to the decompressor. Most recent methods tend to be based on Lempel-Ziv '77 and '78 principles and standardized in recommendations such as V.42 bis for modems and Microsoft Point to Point Compression (MPPC) for compression on point-to-point data links [6]. Much textual data is compressed and decompressed at the application level (e.g., Zip files) rather than by the network elements, and where this has been done, further attempts at compression by the network are usually fruitless. The typical order of magnitude of compression achieved is about 4, but this varies widely according to the nature of source material, ranging from little over 1 for binary program files up to about 15 for text containing numerous spaces, such as some e-mail messages.

Another type of specialized text compression is used for packet protocol headers. This is based primarily on the removal of redundant information and the use of differences in field values between successive packets instead of full values. These are important where the packets are very short, as with speech. Several cases of IP header compression are used in wireless networks and are described in detail in Chapter 7.

These techniques for discrete data do not work for sources with a continuum (or near continuum) range, such as speech and video. The basic approach for these is to carry out a quantization process in which continuum values are fitted into one of a finite range of steps, and a coded step value transmitted instead. This permits a high degree of compression to be achieved, but with a loss of quality in many cases. Some of the more important examples of speech and video compression are discussed in Chapter 10.

The search for optimum quality involves a degree of trade-off between channel coding and source coding because of the finite amount of bandwidth available, as best quality corresponds to minimum compression and maximum error protection.

2.9 Codes in Mobile Phone Networks

Channel codes of all of the above types are used in mobile phone networks, while source coding is normally handled by the end applications or by remote network elements. In first-generation networks, block codes were used for data transmission. In one example CRC correction was employed using a BCH code when the error rate was comparatively low but switched to an RS code of rate 1/2 when reception became bad.

In 3G networks the use of codes is much more sophisticated. In addition to source coding in many applications, there are several characteristic ways in which codes are used in such networks:

- Channel codes;

- Spreading codes;

- Scrambling codes.

The sequence in which the different types of code are used is shown in Figure 2.4.

In this figure, encryption is the action performed by the end application (or in some cases a router), while ciphering refers to the analogous type of process when performed by the CDMA mobile or base station.

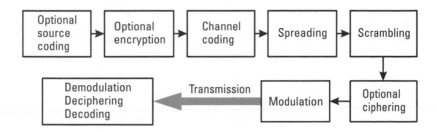

Figure 2.4 CDMA coding sequence.

2.9.1 Channel Codes

These are error-protecting transmission codes of the general types described above, but instead of just using a single code everywhere, a large number of different codes are used according to circumstances. A single user's RF transmissions will normally consist of several distinct transport channels, each of which may in principle use a different coding algorithm appropriate to its own needs. Thus, the channels associated with broadcasts and initial access for WCDMA use convolutional codes of rate 1/2, while the channels associated with data transfer use either convolutional codes of rate 1/2 or 1/3 for low data rates, or a concatenation (turbo) code of rate 1/3 for high data rates with larger block sizes.

The main convolutional codes for both WCDMA and cdma2000 have constraint length of 9.

2.9.2 Walsh Sequences

These sequences are the basis of the orthogonality of spreading codes in CDMA. Although the proper definition is more complex than what follows here, they effectively consist of sequences of binary bits (i.e., 0 or 1) of length equal to an integral power of two, such that the sum of the products of corresponding members of any pairs of sequences is zero. They are also derived by a simple modification from the Simplex codes [4] that are outlined in Section 2.9.4 in relation to scrambling. The earliest version of cdma2000 uses Walsh sequences of length 64 (and period 6) to distinguish channels in the forward direction from the base station to the mobile, with the pilot channel (continuously broadcast by the base station to enable mobiles to identify it) always given the sequence with all zeros. WCDMA uses analogous orthogonal variable spreading factor (OVSF) codes that differ from the Walsh sequences by modifications to improve their autocorrelation properties. The problem of the poor autocorrelation of Walsh sequences is that multipath transmission associated with physical obstacles leads to signals arriving with slightly different delays, thereby making them difficult to distinguish.

2.9.3   Spreading Codes

A spreading code consists of a channelization code and a scrambling code. The channelization code is the characteristic feature of code division multiple access. Instead of transmitting a single symbol, each one is transformed into a number of chips, with the number of chips defining the spreading function, thereby increasing the bandwidth by a corresponding factor.

Suppose that $d(t)$ is a FEC-coded data bit at time t that is signaled over a time interval of T seconds. Then, it is multiplied by the spreading code $s(t)$ to produce a sequence of spreading factor (SF) chips (symbols), each of duration T_c seconds, where $T/T_c = \text{SF}$ and SF is the spreading factor. As a result, the bandwidth required goes up by a factor of SF. The elements of the spreading code $s(t)$ are elements of the Walsh sequences mentioned earlier in this chapter and consist of orthogonal sequences of +1 and −1 (where −1 corresponds to bit value 0) of dimension SF. The receiver multiplies its total incoming signal by the same spreading code in order to extract its own channel as required.

Both WCDMA and cdma2000 operate at fixed chip rates (3.84 Mcps and 1.2288 or 3.6864 Mcps, respectively) and with spreading factors that are multiples of 2 ranging from 4 up to a maximum of 512. The maximum data rate corresponds to the minimum spreading factor.

Cdma2000 uses orthogonal constant spreading factors (OCSFs), whereas WCDMA uses OVSFs that permit SF to be changed during a transmission. This enables WCDMA to handle variable transmission rates by means of the following mechanism.

Suppose the FEC-encoded data rate is $R = 1/T$ as above with spreading factor SF that defines the dimension (length) of a Walsh sequence. The data rate, then, can be increased to a higher rate KR, where $K = 2^{(n-1)}$ and n is a positive integer by mapping n-bit sequences of data into one of the $2^{(n-1)}$ SF-dimensional Walsh codes within T seconds. This is equivalent to mapping each data bit into an SF/K-dimensional Walsh code within a period of T/K seconds.

2.9.4   Scrambling Codes

The second part of the spreading code required to make transmissions on the same frequency distinguishable is the scrambling code, by which the data is also multiplied modulo-2 by the exclusive OR (XOR) operation (but without increase of bandwidth). Scrambling codes are based on pseudonoise (PN) sequences. These are repetitive codes with a period, N, such that they possess the following random-like characteristics:

- The number of zeros and ones in a sequence are equal if N is even and differ by one if N is odd. Thus, the probability of a bit being zero or one is effectively one-half.

- Within a sequence of length, N, runs of characters satisfy the following:

  - Half of all runs have length one, and half of these are zero;

  - A quarter of all runs have length two, with half of these zero;

  - An eighth of all runs have length three, with half of these zero.

- The autocorrelation of the sequence is independent of position in the sequence.

The codes with this property are the Simplex codes $S(N)$ with length $2^N - 1$, dimension N, and minimum distance $2^{(N-1)}$, also known as m-sequences. The very large value of the minimum distance makes them extremely suitable for distinguishing between different transmissions over the same frequency range. The pseudorandom nature of the sequences also means that nonoverlapping subsequences are almost uncorrelated, hence mutually orthogonal and usable themselves for distinguishing between separate transmissions.

The actual codes used are mostly derivatives of the m-sequences that are based on these nonoverlapping subsequences. The length and precise usage of the m-sequences differ both between WCDMA and cdma2000, and between the uplink (UL) from the mobile to base station and the downlink (DL) to the mobile.

In WCDMA there are $2^{18} - 1$ (i.e., 262,143) of these scrambling codes on the DL, divided into 512 subsequence sets consisting of a specific primary scrambling code and 15 secondaries, with the remainder unused. Each cell of the RF network has one and only one primary code, which distinguishes its transmissions from those of its neighbors. If only the primary is used, then all the spreading codes will be orthogonal, but a secondary may be introduced to handle temporary excess users in the cell. If this happens, the codes will no longer be completely orthogonal and there will be some degradation of quality due to interference. On the UL the codes are based on a sequence of length $2^{25} - 1$, but using subsequences of 38,400 chips corresponding to 10-ms duration radio frames on specific channels.

In cdma2000, scrambling uses a code of length $2^{42} - 1$ related to the electronic serial number of the mobile plus a privacy factor to prevent eavesdropping based on a known serial number for the UL. On the DL an m-sequence of length $2^{15} - 1$ is used, where the 32,768 bits are divided into 512 consecutive segments of 64 bits that each identify transmissions from a specific base station.

The function of the scrambling codes also differs according to the direction of transmission, UL or DL. On the UL the scrambling code distinguishes between different mobiles and channels, but on the DL it distinguishes between cells or sectors.

2.9.5 Security Codes

There are three types of security feature in a 3G network:

- Scrambling codes;
- Security masks;
- Network-level encryption via IPSec.

The scrambling code itself forms the first line of security, as without its knowledge an interceptor cannot isolate an individual call. In the case of UMTS, the scrambling code is an arbitrary choice of PN-sequence from the possible range, but for cdma2000 the scrambling code for message transfer is largely determined by masking the PN-sequence with a permutation of the bits belonging to the mobile's 32-bit electronic serial number (ESN) that is in the public domain. The ESN acts as a public code, so security requires an additional private code. The scrambled data transmitted in cdma2000 is formed by masking the sequence scrambled by the public code with the private code via an XOR operation. The scrambling and security code masking apply only to the radio link, so additional security is advisable elsewhere. This is often provided by end-to-end IPSec encryption. Without the use of encryption, many applications are vulnerable to hacking, which leads to denial-of-service attacks. This is particularly true of video applications, where spoofing of call features can cause flooding of the network with vast amounts of video data. These topics are outside the scope of this book, but further information can be found in general books [7], and in 3G specifications [8] and standards [9, 10].

2.9.6 Source Coding and Data Compression

Payload data compression is not part of the 3G standards, as they assume that most such compression will be performed by the end applications. GPRS, however, does support use of V.42 bis and V.44. Extensive support is given to various forms of IP header compression (see Chapter 7 for a description of these and also the chapters on radio technologies).

The principal forms of source coding that the mobile has to perform itself are speech compression and videotelephony compression. The main forms of these are described in Chapter 10.

2.10 Modulation

When coded data is transmitted over a radio link, it has to be sent as a series of modulated pulses. There are two main modulation schemes that are used in 3G networks: binary phase shift key (BPSK) and quadrature phase shift key (QPSK) (there are a number of other schemes used in GSM and GPRS). The essence of BPSK and QPSK is to encode logical 1 and 0 by different phases of the radio signal, and decode by selecting the binary signal corresponding most closely to the detected phase shift. BPSK uses phases of 0 and 180, while QPSK uses 45, 135, 225, and 315. BPSK provides a purely real (in the sense of complex numbers) signal, while that for QPSK also has an imaginary component. An advantage of the latter is that it provides a lower ratio of peak-to-average signal magnitude. In BPSK each pulse represents a single bit, whereas a QPSK pulse represents 2 bits, thereby providing double the information rate.

In the simpler case of BPSK, the modulator receives a stream of inputs, b, with magnitude +1 or −1 (for 1 and 0), each lasting for a time T, and outputs a stream of pulses, $S(t)$, of duration T and amplitude g with phases 0 or 180 at the carrier frequency f_C.

$$S(t) = \mathrm{bg}\cos(f_C t) \tag{2.8}$$

where $b = \cos(0) = 1$ for 1, and $b = \cos(180) = -1$ for 0. The effect of spreading is to replace the bit b by $bc(t)$, where $c(t)$ is a sequence of short binary pulses (or chips) of duration T_c.

In many cases, including CDMA networks, the coded traffic is split into two channels, the in-phase (*I*-channel) and quadrature (*Q*-channel), before modulation. The *I*-branch consists of all odd-numbered bits, and the

Q-branch consists of all the even-numbered bits. Thus, the modulated QPSK signal can be expressed as

$$S(t) = (g/\sqrt{2})I(t)\cos(2\pi f_C t) - \left(g\sqrt{2}\right)Q(t)\sin(2\pi f_C t) \qquad (2.9)$$

where $I(t)$ and $Q(t)$ are the two pulse trains and f_C is the carrier frequency. GPRS uses another scheme, Gaussian minimum shift keying (GMSK), while EGPRS also uses 8-PSK. For details of modulation schemes in general, see [11, 12].

References

[1]   Shannon, C. E., "A Mathematical Theory of Communication," *Bell Syst. Tech. J.*, Vol. 27, 1948, pp. 379–423.

[2]   Hamming, R. W., "Error Detecting and Error Correcting Codes," *Bell Syst. Tech. J.*, Vol. 29, 1950, pp. 147–160.

[3]   Golay, M. J. E., "Notes on Digital Coding," *Proc. IEEE*, Vol. 37, 1949, p. 657.

[4]   Bossert, M., *Channel Coding for Telecommunications*, New York: John Wiley & Sons, 1999.

[5]   Schlegel, C., *Trellis Coding*, New York: IEEE Press, 1997.

[6]   IETF RFC 2118, "Microsoft Point To Point Compression," 1997.

[7]   Kaeo, M., *Designing Network Security*, Indianapolis, IN: MacMillan Technical Publishing, 1999.

[8]   ETSI TS 133 102, "3G Security Integration Guidelines," V.4.2.0 (2001), http://www.etsi.org.

[9]   IETF RFC 2402, "IP Authentication Header," 1998.

[10] IETF RFC 2406, "IP Encapsulating Security Payload (ESP)," 1998.

[11] Vucetic, B., and J. Yuan, *Turbo Codes: Principles and Applications*, Boston, MA: Kluwer Academic Publishers, 2000.

[12] Xiong, F., *Digital Modulation Techniques*, Norwood, MA: Artech House, 2000.

3

WCDMA

3.1 Nature and Principles

WCDMA is the version of DS-CDMA proposed by the 3GPP to meet the needs of the UMTS defined by IMT-2000. Under the banner of WCDMA there are actually two different standards, WCDMA-FDD and WCDMA-TDD. They both use a common network architecture and differ primarily at the physical level. The specifications have evolved through a series of releases. This book is based on Releases 4 and 5 functionality, where the basic distinction between these two is inclusion in the latter of support for the Internet multimedia core (IM) to provide both PS and CS core-network functions (see Section 8.1). The earliest UMTS networks make use of R1999 specifications (the last release before Release 4), which has essentially the same QoS features as Release 4. In this and other chapters, the references are to published ETSI standards in preference to 3GPP draft specifications, but at the time of writing the Release 5 material had not been published in this way, so general 3GPP references are given for these (this applies particularly to Chapter 8).

Apart from code division based on spreading codes, the other main characteristic of WCDMA is the use of power control as the means of controlling quality for a given bit rate. The specifications contain detailed recommendations for the power that should be used for various combinations of error rate and bit rate. Use of excessive power will cause interference problems. The aim is to use the minimum power necessary to achieve a specified level of quality and performance, rather than to try to obtain top quality for an individual transmission.

Unlike GSM or cdma2000, WCDMA does not require precise synchronization. This makes operation within large or overshadowed buildings, where reception of GPS timing signals is obstructed, much easier, but this has the downside of making search and handover procedures harder (see Section 3.4.7).

A UMTS mobile network consists of a UTRAN and a fixed core network; together they constitute a public land mobile network (PLMN). The architecture of the UTRAN is shown in Figure 3.1 and its features are described in this chapter. (The interface to the core network is covered in Chapter 8.)

The main components of the UTRAN are as follows:

- *UE:* This refers to the complete unit of a 3G mobile phone, its SIM card, and any terminal equipment that may be docked onto the phone.
- *Node B:* This is the logical function of the base transceiver station (BTS) concerned primarily with the physical level functions. Higher-level control is directed by the RNC. A single Node B can support several cells.

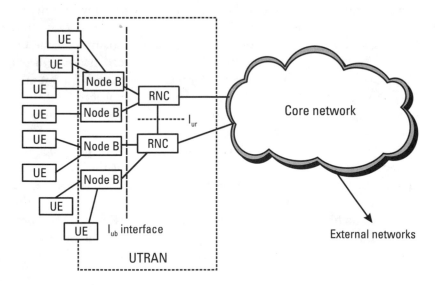

Figure 3.1 UMTS architecture.

- *Radio network controller (RNC):* This is responsible for resource control for one or more Node B, and for providing the interfaces to the core network. In general, the UTRAN will operate in more than one frequency band, each of which represents a different radio network subsystem (RNS) containing at least one Node B and at least one RNC. In order to support handovers of users as they move through the network, the functionality of the RNC is divided into several distinct roles. A single physical RNC supports all the functions, but a single UE may obtain the functions from separate physical RNCs. These roles are as follows.

1. *Controlling RNC (CRNC):* The RNC that controls a given Node B is its CRNC and is responsible for load and congestion control of all its cells on the basis of UL interference and DL power reports from the Node B.

2. *Serving RNC (SRNC):* The SRNC is the RNC that is responsible for handling both the signaling and user traffic for a single UE from the UTRAN to the core network, and also terminates the radio resource control (RRC) signaling between the UE and the UTRAN. It sets the quality target for UL signals that form the basis of outer loop power control (see Section 3.4) and also performs the layer 2 functions (see Section 3.3) for the data transfer between the RNC and the radio link. An SRNC may be the CRNC for the Node B used by the UE at any time, but not necessarily so. Each UE has a single SRNC at any one time.

3. *Drift RNC (DRNC):* Because of mobility a UE may make use of more than one cell and more than one RNC at any time; the DRNC is any other RNC whose cells the UE is communicating with. The DRNC can combine the signals from the separate cells (see Chapter 6 for the benefits of doing so). It does not perform level 2 processing but passes data for dedicated channels transparently from the Node B to RNC interface on to the interface between the distinct physical RNCs.

The interface between a Node B and its CRNC is referred to as the I_{ub} interface, and that between two distinct physical RNCs is the I_{ur} interface. Signaling over the I_{ub} [1] is covered by Node B application part (NBAP) specifications [2], while the radio access network application part (RANAP) [3] governs that over the I_{ur}.

A single physical RNC, Node B, or UE may support both WCDMA-FDD and TDD, but need not do so. The basic distinctions between FDD and TDD are described in the following paragraphs.

3.1.1   WCDMA-FDD

In most parts of the world this will operate in the frequency band 1,920 to 1,980 MHz for the UL from the UE to the base station (usually referred to as Node B in 3GPP documents), and in the band 2,110 to 2,170 MHz for the DL from the base to the user. There will also be a few areas where the corresponding bands are likely to be 1,850 to 1,910 and 1,930 to 1,990 MHz. The spacing between individual transmission channels is a nominal 5 MHz, while the central frequencies must be integral multiples of 200 KHz. Frequency pairing in FDD operation has the result that if traffic volumes are highly asymmetric between UL and DL (e.g., with data streamed to mobiles from hosts on the Internet), then there is liable to be unusable spare capacity in the less heavily loaded direction (the UL in this example). National authorities issue licenses to operators for a portion of this bandwidth in their area of jurisdiction—with 15 MHz a fairly typical allocation—thereby permitting several carrier frequencies each.

3.1.2   WCDMA-TDD

WCDMA-TDD uses the same overall 2-GHz frequency band, but unlike the FDD version, the UL and DL operate at the same frequency. This is achieved by using synchronized time intervals for the two operations. WCDMA-TDD has a wideband option using the same 3.84-Mcps rate as for FDD, but also a narrowband option at 1.28 Mcps. In the wideband version a standard radio frame of 10 ms duration is divided into 15 timeslots on the basis of 2,560 chips per slot. Unlike FDD, TDD can share bandwidth between UL and DL according to demand, so a higher grade of service overall can be provided if the asymmetric applications can be put onto TDD.

WCDMA radio interface has a three-tier architecture roughly corresponding to the open systems interface (OSI) protocol stack, as shown in Table 3.1. The relationship between these functions, and whether they belong to the control- or user-plane, is shown in the protocol architecture diagram of Figure 3.2.

The roles of these layers are described in the following sections.

Table 3.1

WCDMA Protocol Components

Layer	WCDMA Functions
3	RRC
2	Media access control (MAC)
2	Radio link control (RLC)
2	Packet Data Convergence Protocol (PDCP)
2	Broadcast and multicast control (BMC)
1	Physical layer

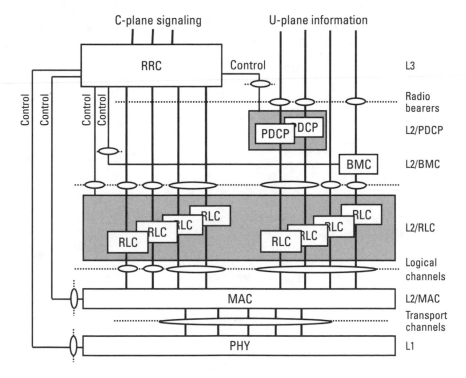

Figure 3.2 Radio protocol architecture. (*Source:* [4] © ETSI 2001.)

3.2   Layer 3 (RRC)

This layer is responsible for radio control in general; in particular, it is responsible for the routing of higher layer messages, broadcast control, paging of mobile devices without an active connection, dedicated control of all functions specific to one mobile, and handling of transfers. It exists in the control-plane and to a lesser extent in the user-plane. There is also a small duplication avoidance sublayer between the RRC and the core network transport. The RRC consists of four functional entities [5] as follows:

- *Routing function entity (RFE):* This handles the routing of higher layer messages to the appropriate UE mobility management or call management entities or the core network functions.

- *Broadcast control function entity (BCFE):* This is responsible for delivering resource control broadcasts, notably system information.

- *Paging notification function entity (PNFE):* This is responsible for paging UEs that do not have resource control connections.

- *Dedicated control function entity (DCFE):* This handles all the control functions that are dedicated to a single UE. In the case of TDD it is supplemented by a shared control function entity (SCFE) that handles UL and DL resources for shared PS channels.

The procedures that affect QoS for a specific UE belong to DCFE and are as follows:

1. Establishment, reconfiguration, and release of radio bearers compatible with requested QoS;
2. Control of transport format combinations;
3. Assignment of access service class priority for use of random access channels;
4. Control of UE capability exchange between UE and UTRAN;
5. Traffic and signal quality measurements;
6. Outer-loop power control;
7. Physical channel allocation for TDD.

Radio bearers (RBs) comprise both signaling radio bearers (SRBs) for resource control messages and radio access bearers (RABs) for the user traffic.

The RRC sets up three or four SRBs, two for resource control messages, one for high-priority network access signaling and optionally another such bearer for lower priority messages. A single call from a UE may make use of radio links to more than one cell, with several RABs on each with different QoS features (e.g., for audio and video in a multimedia application). A request for radio bearer establishment or reconfiguration may contain information elements that are directly related to lower layer functions. These include RLC information for SRBs (see Section 3.3.3), RLC, PDCP (see Section 3.3.1), RB mapping information on transport format sets (see Section 3.4) for RAB, and a quality target for dedicated traffic channels.

The transport format sets specify combinations of data block sizes that can be multiplexed onto transport channels at level 1; PDCP handles data header compression; and UE capability exchange is required to support codec and media negotiation between end points to avoid transcoding and its consequent delays and corruption.

Communication between the Node B and RNC [2] uses ATM as a transport protocol, employing AAL5 cells for control traffic and AAL2 (see Section 7.3) for user traffic. This contributes a few milliseconds to network transfer delays, but otherwise has no significant effect on QoS.

The RRC is logically situated between various network service access points and service access points to the top RLC part of layer 2. Physically it is located in the RNC that controls the Node B logical base stations and the mobile. The RNC is itself normally colocated with one or more Node B; for example, there may be several sectors of a cell being supported from the same location.

Both the UE and Node B (base transceiver) possess an RRC layer with associated states, modes, and processes. The processes are those that entail changes in radio resources, such as initial cell selection or reselection, establishment or reconfiguration of radio bearers (the level 2 services), control of requested QoS, UE quality measurement reporting and control, and outer loop power control. The modes for a UE, when switched on, are either idle or connected, with the latter having several states. In idle mode the UE just looks for higher priority networks. It initiates establishment of a connection to the UTRAN by sending an RRC connection request to the UTRAN. The RRC layer in the RNC performs admission control, assigns radio resource parameters [e.g., transport channel types and formats (see Section 3.3.4)], and establishes radio links from relevant Node Bs if a dedicated channel is involved. If connected to the UTRAN but not in the process of executing a call, the UE will be in either the URA_PCH or CELL_PCH states, essentially listening to paging and broadcast messages, performing measurements, reselecting cells, and looking for higher priority networks. An important QoS procedure that can occur at this stage is the sending of a UE capability information element (IE) to the UTRAN in response to a UE capability inquiry or automatically with the connection request. There are many components to this, but some of the more important for QoS are PDCP capability, physical channel capability (types and speeds), radio frequencies supported, RLC capability (buffer and window sizes), transport channel capability (types, numbers, block sizes, and formats), and timer values (see Section 3.6). In order to execute a call, a radio bearer must be established.

The main state for active calls is CELL_FACH in which the UE has both dedicated control channels (DCCHs) and dedicated traffic channels (DTCHs) available. The RRC functions of the UE in this state are to perform cell reselection as required, carry out measurements, run cell update timers, and listen to all forward access channel (FACH) transport channels that are mapped onto its control channel.

The UTRAN broadcasts a large number of system information blocks via Node B to the UEs. Each of these message blocks has a header with a scope field (e.g., CELL) and a state field (e.g., CELL_FACH), so that a UE updates its system information if the scope and state are appropriate to its condition. There are 18 such blocks, but they are not directly related to QoS features.

One of the main ways in which the RRC controls quality is through the transport combination format set (TFCS) that it passes to the media access control (MAC) layer (see Section 3.3.4). Another essential factor is the monitoring of signal-to-interference ratio (SIR) on a pilot channel. The RRC issues outer power loop control messages on the basis of this information. The RRC communicates with the top of the four level 2 sublayers, radio link control (RLC) via the radio bearers, which are separate channels depending on the type of information sent (see Figure 3.2).

3.3   Level 2

3.3.1   Packet Data Convergence Protocol

The Packet Data Convergence Protocol (PDCP) [6] is another level 2 function that provides services to the nonaccess stratum (NAS) at the UE or relays at the RNC. Its main role is in control of header compression for IP packet data. This is important for QoS for any real-time traffic that uses small IP packets (e.g., speech) because the headers may be several times as long as the actual user traffic and hence increase by a large factor both the transmission time and capacity used. For example, VoIP is often carried in RTP/UDP/IP where the uncompressed header is 40-bytes long, while the speech typically only occupies 20 to 24 bytes (depending on the voice codec), but compression can cut the header to only 4 bytes most of the time. Header compression is not helpful for large packets and, hence, is not often used by file transfer applications.

Several distinct header compression protocols can be used, and PDCP inserts a protocol identifier field (PID) in the protocol data unit (PDU) to indicate the type, if any, to be used. This is a 4-bit field, with "0" indicating no header compression, and most of the others being options defined within the Internet RFC2507 and RFC3095 on compression. RFC3095 (represented by PID 9–11) is the most resilient method on error-prone links such as the radio link. The significance of these options is described in Section 7.4.10, where the features and benefits of the protocols are described. No method is currently defined for SIP (see Section 7.4. 8) header compression, and this may have to be added later.

3.3.2 Broadcast and Multicast Control

There is one broadcast and multicast control (BMC) entity per cell in the UTRAN [7]. It communicates with the RLC by means of the UM mode. Its functions include the following: storage of cell broadcast messages, traffic volume and radio resource information to CBS, scheduling of BMC messages to UE, delivery of BMC to UE, and delivery of CBS messages to NAS. There is also an extension to this layer that handles multimedia multicasts [8].

The traffic volume information is an estimate of the current cell broadcast service data rate requirement in kilobits per second; this is sent to the RRC.

3.3.3 RLC

The RLC [9] sits between the level 3 RRC and the MAC layer. It has three modes of operation—transparent mode (TM), unacknowledged mode (UM), and acknowledged mode (AM)—that are directly relevant to QoS. Transparent mode takes data to or from the RRC and segments or reassembles it, but does not add other header information; it is used for delay-sensitive traffic that is tolerant of errors. Unacknowledged mode also performs these functions, but adds or checks sequence numbers for the MAC layer and may add ciphering if needed; it is used for traffic that requires delivery in sequence and does not require delivery of errored packets. The acknowledged service is for data that has a low error tolerance and includes addition of a CRC field, flow control, and error checking to the UM functions, with ARQ retransmission when errors are found. AM guarantees delivery of data, but either with or without sequencing.

The RRC uses different radio bearers for signaling and user traffic. Those for signaling are shown in Table 3.2, which shows the control channel (see Section 3.3.4) supported and the RLC mode used. The NAS above posts user traffic to the RLC, bypassing the RRC, but possibly going through the PDCP (see Section 3.3.1) or BMC (see Section 3.3.2) layers. The bearer configuration is determined by the RRC and is defined in terms of static, semistatic, and dynamic parameters determined by the required QoS. The static part includes the RLC information, whose main features are the choice of mode, whether sequencing is required, the permitted RLC PDU sizes, the acknowledgment timer, and the transmit/receive window sizes in terms of RLC PDUs. The semistatic and dynamic parts relate primarily to the transport formats specified for MAC to L1 operation. RLC segmentation and reassembly fits data from higher or lower levels into the permitted PDU sizes, with an extension flag and length indicator after the sequence number at the start of the PDU if several higher level SDUs are fitted into a single PDU. There is one such indicator per SDU, each indicating the number of octets in the SDU.

The majority of quality control features in RLC, therefore, lie in the use of acknowledged mode. Flow control is based on a 12-bit sequence number, window sizes, and an implementation-dependent superfield for

Table 3.2
Use of Radio Bearers for Signaling

RB Number	Services Carried
0	All RRC messages on common control channel (CCCH)
	RLC-TM on UL, RLC-UM on DL
1	Messages on DCCH when using RLC-UM
2	Messages on DCCH when using RLC-AM
3	Messages on DCCH RLC-AM that carry higher-level signals
4	Optionally the same as 3
5–31	Messages on DCCH using RLC-TM

Source: [5].

selective acknowledgments of received and missing PDUs. An error burst indicator can be used to show that the next codeword received in the super-field will indicate the number of subsequent PDUs that are erroneous. The RLC also receives information from the RRC specifying the maximum number of retransmissions permitted for a PDU and how often the transmitter should poll the receiver in the case of a PDU or SDU containing a poll bit.

The RLC communicates with the MAC layer via logical channels that depend on the function of the information; these are more detailed than the radio bearer selections above. AM is only applicable to two types of logical channel: DCCH and DTCH.

3.3.4 MAC Layer

The MAC configuration is controlled by the RRC, and the MAC [10] uses logical channels to communicate with the RLC and transport channels to communicate with the physical layer. The logical channels consist of control channels and traffic channels as follows:

- *Control channels:* Broadcast (BCCH), paging (PCCH), DCCH, CCCH, and shared (SHCCH);

- *Traffic channels:* DTCH and common transport channel (CTCH).

The transport channel types, each of which is defined unidirectionally, include random access channels (RACHs), FACHs, DL shared channels (DSCHs), common packet channels (CPCHs) for FDD on UL only, UL shared channels (USCHs) for TDD only, broadcast channels (BCHs), paging channels (PCHs), and dedicated transport channels (DCHs). Each RLC radio bearer can use up to two logical channels, where one is for control traffic and the other for user traffic.

The MAC layer has three functional entities, as shown in Table 3.3. Each UE has one MAC-b, while the UTRAN has one MAC-b per cell. There is one MAC-c/sh in the UE that handles the transport-channel-type field (TCTF) header field in the MAC to indicate common logical channel type for a common channel. For dedicated logical channels, it handles UE identification, transport format (TF) selection, and TF combination (TFC) selection on the basis of TFC set configured by the RRC. There is one MAC-c/sh in the UTRAN that is located in the CRNC. It performs priority

Table 3.3
MAC Functional Entities

Entity	Function
MAC-b	Handles the BCH
MAC-d	Handles DCHs
MAC-c/sh	Handles the remainder, notably CTCHs

handling via the access service class for the FACH and DSCH, TCTF multiplexing, TFC selection, and DL code allocation for use on DSCH.

The MAC-d in the UE performs transport channel switching on the basis of decisions by the RRC, multiplexing of multiple dedicated logical channels onto a single transport channel, TFC selection on the UL, and ciphering/deciphering for RLC-TM if required. The MAC-d for the UTRAN is situated in the SRNC. Its functions are similar to those for the UE but with the addition of priority setting on data received from the DCCH or DTCH, DL priority and schedule handling of transport channels allowed by RRC, and flow control towards the MAC-c/sh in order to limit level 2 buffering and latency.

For peer-to-peer communication, a MAC-PDU consists of a bit string whose length is not necessarily a multiple of eight. It consists of a MAC-SDU plus an optional header, where the MAC-SDU is either a bit string of any nonzero length or an integer multiple of octets and the header is of variable size. The header consists of the following:

- TCTF of variable size to identify the logical channel type being used over a FACH or RACH transport channel (e.g., BCCH, CCCH, CTCH, SHCCH, or dedicated logical channel);

- A UE identity type and UE identity;

- A C/T field that indicates the instance (1–15) of a logical channel when multiple logical channels of the same type are carried over a single transport channel.

When a DTCH or DCCH is mapped onto a DCH without any multiplexing of dedicated channels, no MAC header is required; if there is any such multiplexing, then the header consists of the C/T field. Numerous other situations require other headers; one that requires each of the above fields is the mapping of a DTCH or DCCH to RACH or FACH.

The UE identity type can either be an UTRAN radio network temporary identity (U-RNTI) with 32-bit identifier, or a cell radio network temporary identity (C-RNTI) of 16 bits.

The MAC receives configuration information from the RRC in MAC-Config-REQs that specify the priority, 1 to 8, to assign to each logical channel carried on a radio bearer when the latter is set up or reconfigured. This priority is used to select the TFC in the UE.

The manner in which different logical channels can be combined into transport channels for MAC to L1 communication is determined by the transport formats [11] controlled by the RRC layer. The basic entities involved in this are as follows:

- *Transport block (TB):* This is the basic unit exchanged between the MAC and level 1 and corresponds to a MAC PDU.

- *Transport block set (TBS):* This is a set of TB that can be exchanged between the MAC and L1 at the same time and on the same transport channel. All blocks in a TBS must have the same size.

- *Transport block size:* This is the number of bits within a TB. It is set to accommodate the RLC size plus any appropriate MAC header. Permissible sizes are 1 to 5,000 bits.

- *Transport block set size:* This is the number of bits in a TBS.

- *Transmission time interval (TTI):* This is a key parameter for QoS that is defined as the interarrival time for TBS. It is a multiple of the 10-ms radio frame duration, with possible values of 10, 20, 40, or 80 ms.

- *TF:* This determines the format for TBS exchange between L1 and MAC and consists of dynamic and semistatic parts. The dynamic attributes are transport block size and transport block set size for both FDD and TDD, plus TTI in the case of TDD only. The semistatic attributes are the TTI for FDD (and optionally for non-real-time TDD), the error protection scheme to use, and the CRC size. The error protection scheme consists of the type of channel coding to use, if any (i.e., convolution or turbo), the coding rate, and the static rate matching parameter. TTI and the TB set size together determine the instantaneous bit rate of the transport channel.

- *TFS:* A single transport channel may be permitted to use more than one TF, provided that they all have the same semistatic part and the set of permissible values defines the TFS.

- *TFCS:* Level 1 can multiplex several transport channels together to form a single physical channel (see Section 3.4), in which case each transport channel has its own TFS, but only certain combinations of these (the TFCS) are allowed.

- *Transport format indicator (TFI):* This is an index used by L1 and MAC to identify a specific TF in the TFS.

- *Transport format combination indicator (TFCI):* This is an index used to identify one particular TFC from the TFCS. On the basis of the radio priorities, the MAC makes the choice of TFC for the available TB.

Some applications (e.g., multimedia) may use several distinct traffic streams, each with its own block sizes and QoS needs. In that case they are likely to use more than one logical channel and transport channel, each with its own transport formats. Some may also have variable bit rates (e.g., AMR voice where the degree of compression and type of coding varies according to signal quality), and this is achieved by having several possible block sizes in the dynamic part of the TFS. A formal example of this is shown in Figure 3.3.

This diagram shows two transport channels (TrCHs): TrCH (1) and TrCH (2), with TTI (1) and TTI (2), respectively. TrCH (1) has only one block size, TB (1), but has a block set of either one or two blocks depending on the particular transmission interval; whereas TrCH (2) has a block set of one, but a TFS that includes the two sizes for TB (2) and TB (3). In the case

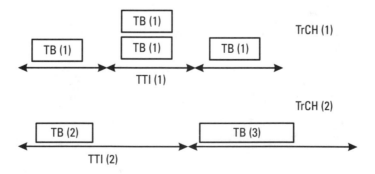

Figure 3.3 Transport formats.

of TrCH (2), a TFI is needed to indicate which size is in use for any given TTI period.

The dynamic part of the TF in FDD consists of the block size and block set size; so in this diagram the TFS for TrCH (1) has two formats, $\{x,x\}$ and $\{x,2x\}$, where x denotes the block size of TB (1). TrCH (2) also has two possible formats with dynamic parts, $\{y,y\}$ and $\{z,z\}$, where y and z are the block sizes of TB (2) and TB (3), respectively, and z is greater than y.

In general, several transport channels are likely to be multiplexed onto a single CCTrCH, and level 3 then dictates the combinations of TF that may be used. One of the main criteria used is to create TFCs that permit the MAC to perform simple rate control without involving any signaling to level 3. This is achieved by specifying combinations that tend to average the over-all rate through combining low throughput options for some channels with higher rate formats for others and by forbidding combinations of extreme formats. Thus, in the case of the very simple example in Figure 3.3, the TF (1) with dynamic part $\{x,x\}$ would be combined with the TF (2) with dynamic part $\{z,z\}$, while $\{x,2x\}$ would go with $\{y,y\}$, thereby giving a TFCS with these two TFC. The choice of TFC can vary on a per radio frame basis and is usually indicated by a 10-bit TFCI (coded for resilience to 32 bits) in the frame, but with blind format detection (based on either CRC or power analysis) as an alternative for relatively fixed rates.

One of the simplest possible real examples would be the case of a speech call using GSM-EFR (see Section 10.1.1). This uses three classes of speech bit with different error protection requirements, so that they are sent on separate transport channels. Each uses a TTI of 20 ms and a block set consisting of a single block, but with block sizes of 81, 103, and 60 bits, respectively. There is only one TF for each channel with respective dynamic parts, $\{81,81\}$, $\{103,103\}$, and $\{60,60\}$, while the semistatic parts of the three TF each have TTI of 20 ms, but differing coding parameters. On the FDD UL (and often on the DL) these would be combined into the same CCTrCH with a single trivial TFC. The default speech codec for UMTS is UMTS-AMR, which has multiple formats for each of three channels, and this is discussed in more detail in Chapter 10.

The MAC layer performs scheduling via status indications to the RLC. These indicate the number and size of PDUs that the RLC can send in the TTI for each logical channel. The MAC also uses the priorities assigned to the logical channels by the RRC to decide which TFC combination to use from the TFCS, and it must use that which allows transmission of the most high priority data. The MAC for the UE must estimate the power required for the TFC and only use those that are within its maximum power capability. This selection takes place at the start of the shortest TTI for any member of the TFCS.

The MAC sends traffic measurements directly to the RRC for dynamic radio bearer control. These consist of buffer occupancies for RLC entities and are made both periodically and on a threshold alarm basis.

3.4 Level 1—The Physical Layer

3.4.1 General

The physical level communicates with its own MAC layer via the transport channels and with its peer physical layer via physical channels. Layers 2 and 3 are very similar for WCDMA-FDD and TDD, but there are major differences at layer 1, and unless otherwise indicated, the following refers to FDD.

A physical channel is defined as a code or set of codes in WCDMA-FDD and additionally by a sequence of timeslots in WCDMA-TDD. The information rate of a physical channel varies with the symbol rate derived from the chip rate of 3.84 Mcps and the spreading factor (SF). SF varies from 256 to 4 in FDD on the UL and from 512 to 4 on the FDD DL, while in TDD it varies from 16 to 1. These give respective symbol rates from 15k to 960k symbols/s on FDD UL and 7.5k to 960k for FDD DL, with 3.84M symbols/s to 240 k/s in TDD in either direction.

Spreading codes (as outlined in the previous chapter) are used to distinguish between the following:

- Different channels from the same source via channelization codes;
- Different cells via scrambling codes;
- Different UE via another set of scrambling codes.

One of the main factors influencing QoS is the mapping of transport channels to physical channels.

3.4.2 Physical Channels

Layer 1 maps the various transport channels to/from the MAC layer to specific physical channels that are transmitted at level 1 on the UL and DL. The way in which these channels are mapped is shown in Figure 3.4 for FDD. Physical channels without any transport association are purely for layer 1 functionality between the UE and Node B.

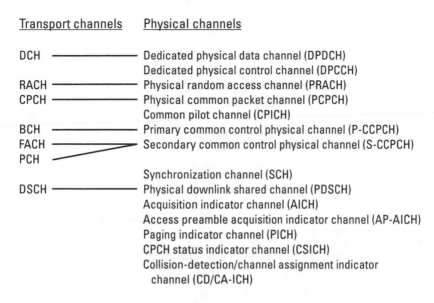

Transport channels   Physical channels

DCH ———————————— Dedicated physical data channel (DPDCH)
                   Dedicated physical control channel (DPCCH)
RACH ——————————— Physical random access channel (PRACH)
CPCH ——————————— Physical common packet channel (PCPCH)
                   Common pilot channel (CPICH)
BCH ———————————— Primary common control physical channel (P-CCPCH)
FACH ——————————— Secondary common control physical channel (S-CCPCH)
PCH ———————

                   Synchronization channel (SCH)
DSCH ——————————— Physical downlink shared channel (PDSCH)
                   Acquisition indicator channel (AICH)
                   Access preamble acquisition indicator channel (AP-AICH)
                   Paging indicator channel (PICH)
                   CPCH status indicator channel (CSICH)
                   Collision-detection/channel assignment indicator
                     channel (CD/CA-ICH)

Figure 3.4 Transport and physical channels. (*Source:* [12] © ETSI 2001.)

The role of many physical channels is outside the scope of this book, and only a subset with above average significance for quality is described below. Physical channels are characterized by means of a specific carrier frequency, scrambling code, channelization code (optional, but used on many), stop and start times in relation to chips, and for the UL only a relative phase. Radio frames correspond to a duration of 15 slots of 2,560 chips each or 38,400 chips in all, hence a duration of 10 ms at 3.84 Mcps.

3.4.2.1 Dedicated Physical Channels

The channels most directly relevant to the transmission of user data are the dedicated physical channels, DPDCH and dedicated physical control channel (DPCCH). CS and most real-time traffic use dedicated channels, while intermittent data traffic can use common or shared channels. The UL DPDCH carries the DCH transport channel from the MAC, while the control channel (DPCCH) is used purely for control information generated in layer 1. In either case each 10-ms radio frame is split into 15 slots of 2,560 chips corresponding to single power cycles, that for DPDCH containing data, while the DPCCH slots carry control information, as shown in Figure 3.5.

The pilot field bits conform to one of a specified set of preset patterns and are used in frame synchronization. TFCI is used for variable rate channels or for channels carrying multiple services, but not for single fixed rate

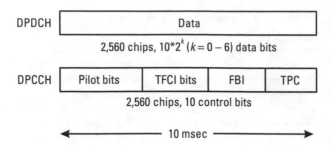

Figure 3.5 UL DPCH slot. (*Source:* [12] © ETSI 2001.)

channels—the decision on TFCI usage is determined by the UTRAN, not the UE. The feedback information (FBI) field is used in relation to diversity [13] and power control during soft handovers. Transmit power control (TPC) indicates whether to change transmitter power level (see Section 3.4.6).

On the DL there is only one type of dedicated physical channel, the DPCH, and it carries multiplexed DPDCH and DPCCH traffic. A single slot contains two data segments and two control segments, as in Figure 3.6.

The two DPDCH segments have unequal sizes that are specified in tables and are dependent on the spreading factor and control TFCI [12], while the parameter k is related to the spreading factor by

$$SF = 512/2^k$$

The DPCCH information consists of TFCI, where applicable, TPC as on UL, and another set of pilot bits with numbers as per the tables. It is possible

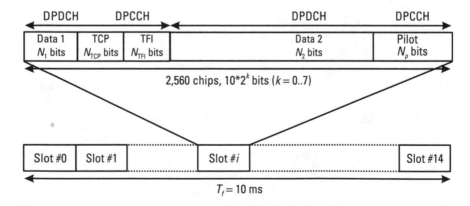

Figure 3.6 DL DPCH slot structure. (*Source:* [12] © ETSI 2001.)

to have more than one DPDCH if the required rate is too high for a single channel. If so, then the spreading factor is 4 for each such channel.

Other physical channels have different structures for their slots [12].

3.4.2.2 Common Physical Channels

The UL random access channel carries 4,096 chip preambles and a 10-ms message radio frame slit into 15 slots, each of which has both data and control parts.

On the DL there are two types of common pilot channel (CPICH) with a fixed rate of 30 Kbps: a primary that is common to an entire cell and defines a phase reference for all other DL channels, except possibly the DPCH, and a secondary for part or all of a cell that gives the phase reference for the DPCH in the exceptional case.

The synchronization channel (SCH) is a DL signal that the UE uses for cell search. It consists of primary and secondary subchannels, so that each slot contains a primary synchronization code of length 256 common to an entire cell, together with a secondary code characterizing a scrambling code in the range 0 to 63 and a slot number in the range of 0 to 14, corresponding to the slot position in the 15 slot subdivision of the 10-ms radio frame.

3.4.3 Physical-Level Procedures

The main functions of the physical level [14] are as follows:

- Error detection on transport channels and indication to higher layers;
- FEC encoding/decoding and interleaving/deinterleaving of transport channels;
- Multiplexing of transport channels and demultiplexing of coded composite transport channels;
- Rate matching;
- Mapping of coded composite transport channels on physical channels;
- Power weighting and combining of physical channels;
- Modulation and spreading/demodulation and despreading of physical channels;
- Frequency and time (chip, bit, slot, frame) synchronization;

- Measurements and indication to higher layers [e.g., frame error rate (FER), SIR, interference power, transmit power];
- Closed-loop power control;
- Combining signals for soft handover;
- Radio frame processing.

3.4.4 Multiplexing, Channel Coding, and Interleaving of Transport Channels

This group of functions [15] prepares the transport channels for transmission over the physical channels and have the greatest relationship to QoS. The main steps in this are as follows:

1. Addition of CRC field to transport block;
2. Concatenation and segmentation;
3. Channel coding;
4. Radio frame equalization;
5. Rate matching;
6. Interleaving;
7. Radio frame segmentation;
8. Multiplexing of transport channels;
9. Physical channel segmentation;
10. Mapping to physical channels.

The size of the CRC check field to add to each transport channel block is decided by higher layers.

All transport blocks to be sent on a specific transport channel within the TTI are concatenated prior to coding. If the resultant code block exceeds the maximum bit size permitted for the type of channel coding to be used, then it is first segmented. The limit sizes are 504 bits for convolution coding and 5,114 for turbo coding, with no limit if no channel coding is required. Specific block sizes are recommended that depend on the data rate and type of transport channel, with turbo coding being used for the highest data rates. The possible code rates are 1/2 or 1/3 for convolutional and 1/3 for turbo. These fractions determine the proportion of transport block bits (as opposed to FEC bits) on the radio link.

Radio frame equalization is used on the UL to ensure that the number of frames is equal to the integral value, F, specified for that transport channel.

This requirement is automatically met on the DL. The technique of equalization is to pad the bit sequence until this criterion is met.

Specific interleaving patterns are used depending on the TTI, whose possible values are 10, 20, 40, or 80 ms, with the complexity and resultant degree of error dispersal increasing with TTI. Where TTI exceeds 10 ms, the radio frames are also segmented into the required number F of frames.

Bits may have to be repeated or punctured in order to ensure that the multiplexed transport channels exactly match the bit rate of the physical channel. Higher layers using detailed algorithms control this.

The resultant frames are then multiplexed into a coded composite transport channel (CCTrCH). WCDMA-FDD has only one CCTrCH on the UL, but can have multiple such channels on the DL. Transport channels that are multiplexed onto the CCTrCH should have comparable QoS needs and, more specifically, the same C/I requirement. Where multiple CCTrCH are in use, each can have its own TFCI but not always, since blind detection of format is sometimes used for each 10-ms frame with fixed rate services. The RACH, CPCH, and BCH are always mapped 1:1 to the corresponding physical channel, but other transport channels may be multiplexed in this way. Dedicated and common transport channels cannot be mixed, so on the FDD UL there is both a dedicated channel CCTrCH and a common channel CCTrCH, of which only one can be active at any instant. The number of transport channels that can be multiplexed onto a single CCTrCH depends on the capability of the UE and on the TTI. This multiplexing is based on taking 10-ms frames from each transport channel involved per TTI. The DPCCH is multiplexed with one of the DPDCH.

More than one physical channel is sometimes used for a single CCTrCH. In that case the CCTrCH is segmented between them by means of a demultiplexing unit, and each of the physical channels must have the same spreading factor. After more interleaving the segments are ready for transmission.

An example of these multiplexing and interleaving procedures is shown in Figure 3.7 for the case of the 12.2-Kbps reference measurements [16] using the transport parameters in Table 3.4.

Table 3.4 and Figure 3.7 show how the DPDCH and DPCCH are multiplexed/demultiplexed and interleaved on a DL DPCH at a rate of 30 k symbols/s for reference measurements. The data rate for the DTCH is the transport block set size divided by the TTI, hence 12.2 Kbps as indicated, which is the typical rate for a speech channel. The CRC (whose size is specified for the TrCH by the RNC) is added to the transport block set, followed by a set of tail bits to enable the coder to determine the code-block boundary.

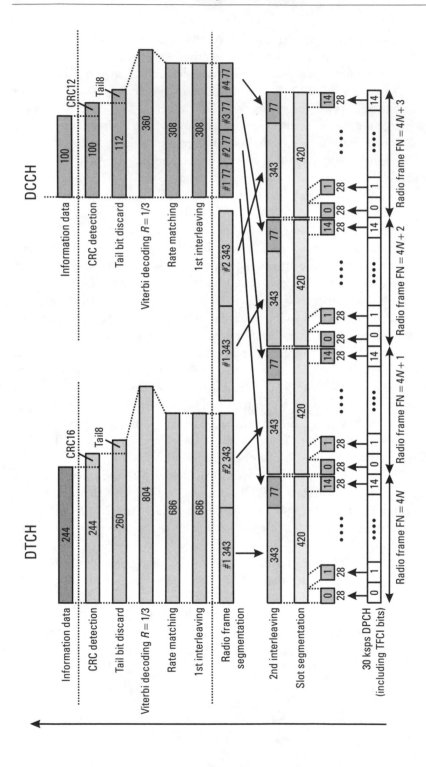

Figure 3.7 Multiplexing/demultiplexing. (*Source:* [16] © ETSI 2001.)

Table 3.4

DL Transport Parameters for 12.2-Kbps Reference Measurements

Parameter	DTCH	DCCH
Transport channel number	1	2
Transport block size (bits)	244	100
Transport block set size	244	100
TTI (ms)	20	40
Type of error protection code	Convolution	Convolution
Coding rate	1/3	1/3
Rate matching attribute	256	256
Size of CRC (bits)	16	12
Position of Tr channel in frame	Fixed	Fixed

Source: [16] © ETSI 2001.

On the DL, higher layer scheduling through the rate matching attribute and resultant padding or puncturing automatically controls rate matching. Each of the 15 slots within a 10-ms radio frame has 2,560 chips, corresponding to 10×2^k bits, where $512/2^k$ is the spreading factor, as in Figure 3.6, where the numbers of bits of various types (e.g., power control and TFCI) within the DPCCH and DPDCH are given by tables [12]. The first interleaving is intrinsic to the separate channels here, and radio frame segmentation is required because the TTI exceeds 10 ms in both channels. The second interleaving combines the first two DCCH segments with the two DPCH segments, and the last two with an additional two DPCH segments for their next TTI. Slot segmentation then splits each of the interleaved segments into 15 slots.

3.4.5 Physical Layer Measurements

In order to achieve the necessary quality, the physical layer has to perform measurements of such features as the block error rate, the SIR, and interference power, all of which must be reported to higher levels so that the RRC in particular can take appropriate action [17]. Maintenance of quality depends on the ability of the system to monitor signal strengths, error rates, and timings, and this is carried out both by the UE and by the UTRAN.

The main measurement made by the UE is the block error rate (BLER) based on the number of CRC errors received.

The UTRAN performs a more complex set of tasks. These include the measurement of the following:

- SIR according to the formula SIR = (RSCP/ISCP) × SF, where SF is the spreading factor, RSCP the received signal code power, and ISCP the interference code power;

- Transmitted power on one channel code on one scrambling code on one carrier;

- Transport channel BER for the DPDCH;

- Physical channel BER for DPCCH;

- Round-trip time (RTT) where RTT = $T_{RX} - T_{rx}$. T_{rx} is the time of transmission of the beginning of the DPCH to the UE, and T_{RX} is the time of reception of the beginning of the corresponding DPCCH/DPDCH from the UE;

- Propagation delay on the PRACH.

In addition to these periodic measures, there are also several sets of measurements to be made at specific stages of operation:

1. Handover measures to compare the relative signal strengths of different cells and, for FDD, the timing relation between cells for support of the soft asynchronous handover procedure;

2. Specific measures for handover to GSM 900/1,800 if relevant;

3. Measures for the UE before random access process;

4. Measures for dynamic channel allocation in TDD.

Some of these measurements have to be made on a different carrier frequency, which can only be made by many UE if DL transmissions are interrupted. In order to enable this, a specific way of operating, called compressed mode, is used. It is implemented by halving the spreading factor (and so doubling the data rate) and using a different scrambling offset by a standard amount from the normal.

3.4.6 Power Control

Power control [13] is a vital feature of WCDMA for limiting interference and controlling quality. As mentioned above, there are several distinct power control loops. The initial power level of the UE is set on the basis of open loop power control. For the UL this entails the UE setting a value for the random access channel based on broadcast information for the cell, while initial strength on the DL is based on measurements by the UE.

The power used on the dedicated physical channels of a call in progress is based on outer and inner loop power control. Outer loop control sets the long-term quality target for the inner loop. Outer loop control resides in the SRNC for the UL and sets the target quality for the inner loop based on measurements of the transport channel quality. In the case of the DL, the function is contained in the UE, but based on targets sent by the SRNC.

Inner loop power control handles the very rapid changes in power that have to be made as a result of the motion of the UE (see Chapter 6), and the function is located in both the UE and Node B. Each serving cell must monitor the quality of signals received on the UL DPCH. The SIR is estimated and compared to the target value. On the basis of this measurement, the serving cell must send a TPC command for every slot (i.e., 1,500 times per second) to the UE indicating 0 if SIR exceeds the target value or 1 if it does not. The UE should increase its transmit power for the UL on receipt of TPC 1 by either 1 dB or a specified TPC_StepSize depending on which of two algorithms is in use.

The UE can also send TPC commands on the UL dedicated physical control channel to the Node B on the basis of its comparison of received quality with the target for adjustment of the DL power.

The power on the physical channel depends on the data rate, so it has to be changed if the TFC is altered. Whenever compressed mode is used to enable the UE to make measurements on the DL, the data rate is increased and hence the power has to be changed.

3.4.7 Cell Search and Handover

One of the most critical stages in a call for QoS problems is at handover. Cell search is required at registration and as the UE moves through the network. During the cell search, the UE searches for a suitable cell based on pilot channel signal strengths [18] and determines the DL scrambling code and frame synchronization of that cell. The cell search [19] is typically carried out in three steps:

1. *Slot synchronization:* During the first step of the cell search procedure, the UE uses the SCH's primary synchronization code to acquire slot synchronization to a cell.

2. *Frame synchronization and code-group identification:* During the second step of the cell search procedure, the UE uses the SCH's secondary synchronization code to find frame synchronization and identify the code group of the cell found in the first step. This is done by correlating the received signal with all possible secondary synchronization code sequences and identifying the maximum correlation value. Since the cyclic shifts of the sequences are unique, both the code group and the frame synchronization are determined.

3. *Scrambling-code identification:* During the third and last step of the cell search procedure, the UE determines the exact primary scrambling code used by the found cell. The primary scrambling code is typically identified through symbol-by-symbol correlation over the CPICH with all codes within the code group identified in the second step. After the primary scrambling code has been identified, the primary common control physical channel (CCPCH) can be detected and the system- and cell-specific BCH information can be read.

If the UE has received information about which scrambling codes to search for, steps 2 and 3 above can be simplified.

Reselection

According to its capability, the UE may monitor a list up to 32 cells on the original frequency when in idle mode, 32 cells on up to two other FDD frequencies, 32 GSM cells, and, if TDD is supported, up to 32 more on up to three frequencies. Once in connected mode the UE has to be capable of maintaining at least six radio links in an active set of cells [20]. The UE performs regular measurements of signal quality and camps on to another cell if signal quality requires this, as indicated by an active set update message from the UTRAN. Handovers can be soft/softer for FDD-FDD on the same frequency, or hard for FDD-FDD at a different frequency, FDD-TDD and FDD-GSM.

The soft handovers do not entail any interruption to data flow. For the others, however, there is an interval between the last TTI containing a transport block on the old DPDCH and the time the UE starts sending a block on the new DPDCH. This interval depends on the number of cells and whether they are already known to the UE, as well as on the type of hard handover. In most cases, the delay is not long enough to trigger a time-out and retransmission of application level data, but this can sometimes occur. For hard FDD-FDD handovers [e.g., between different layers in a cell hierarchy (see Chapter 6)] the formula for the maximum interval without compressed mode is

$$T = T_{IU} + 40 + 20 \times KC + 150 \times OC \quad ms$$

where T_{IU} is not more than one 10-ms frame, 40 represents a channel measurement time, KC is the number of known cells, and OC is the number of unknown target cells in the active cell update message. A cell is known if the UE either had it in its previous active cell list or had made measurements on it in the previous 5 seconds. The factor of 20 on KC increases to 50 if compressed mode is used to support measurements on a different frequency. If the UE only supports six radio links, then this interruption will always be less than 1 second. FDD-GSM handovers entail much shorter interruptions than these maxima and should not cause disruption to service.

3.4.8 Physical-Level Differences for TDD

UL and DL operate on the same frequency in TDD mode, and a TDD channel is a burst that is transmitted within one of the 15 timeslots. There are two possible chip rates: 3.84 Mcps (as in FDD) and 1.28 Mcps. The framing structure differs for the 1.28-Mcps version, and each 10-ms frame is composed of two 5-ms subframes, each of which contains seven normal timeslots and three special timeslots [21].

A burst for 3.84-Mcps TDD consists of two data parts, a midamble, and a guard period. The data parts are spread by channelization and scrambling codes, where the OVSF for the former allows spreading factors of 1, 2, 4, 8, or 16, of which the DL only uses the value 16. Each timeslot can be allocated to either direction, subject to there being at least one timeslot in either direction. The guard period is required to prevent conflicts due to the changes in direction. The midamble is used for training. There are three different types of burst, each with differing ratios of these field sizes. The quantity of transport level data carried by a burst with spreading factor of 1 is 1,952, 2,208, or 1,856 bits according to the type, 1, 2, or 3, and reduced pro rata by the spreading factor for higher SF [21]. In each case the two parts of the data are separated by the midamble, with the guard period at the end of the slot. The basic characteristics of the burst types are as follows:

- Type 1 is suitable for either UL or DL and has a relatively long midamble of 512 chips with a guard period of 96.
- Type 2 is also suitable for both UL and DL and has a longer data field at the expense of the midamble (reduced to 256 chips) with a guard period of 96.
- Type 3 is only suitable for UL and has a midamble of 512 chips and a guard period of 192 chips. The extended guard period makes it suitable for initial cell access or handovers.

The TFCI and TPC commands are included in the data portion of the burst, with the TFCI split into two parts on either side of the midamble and the TPC for the UL ahead of the second part of the TFCI. This slot and burst structure for 3.84-Mcps TDD is shown in Figure 3.8.

In the case of 1.28-Mcps TDD, a 10-ms frame consists of two 5-ms (6,400 chip) subframes that can each be split into seven timeslots, of which the first is always for the DL and the last for UL, with the remainder switchable, subject to two switching points in the subframe. Each timeslot is 864 chips long, and the remaining 352 include DL and UL pilot slots and a guard period for the switching points. Bursts for 1.28 Mcps again have two data fields, a midamble, and another guard interval. For SF of 1 the data field

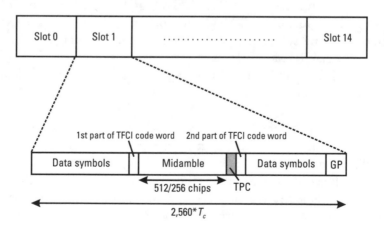

Figure 3.8 TDD slot structure. (*Source:* [21] © ETSI 2001.)

is 352 chips, the midamble 144, and the guard 16, with pro rata reductions for higher SF.

A major consequence of the ability to use slots in either direction is the existence of a UL shared channel, and the possibility to have more than one CCTrCH on the UL as well as the DL.

3.5 High-Speed Downlink Packet Access

High-speed downlink packet access (HSDPA) was introduced in UMTS to provide high throughput PS traffic with low delays and very high peak rates [22]. It entails a new module for the MAC layer, MAC-hs, and a new transport channel, the high speed downlink shared channel (HS-DSCH). It is supported in both FDD and TTD modes. The architecture for HSDPA is shown in Figure 3.9

In this diagram, the frame protocol (HS-DSCH FP) handles the data transport from SRNC to CRNC and from CRNC to Node B, with the MAC-hs situated below the MAC-sh in the CRNC. There is also an alternative mode in which the SRNC is connected directly to Node B, bypassing the CRCN, so that the MAC-sh is not involved.

At RLC level, it only uses AM and UM, rather than TM (omitted on the account of ciphering issues) that normally supports the highest data rates. PDCP can be used as before. It is aimed primarily at the streaming, interactive, and background class applications (see Chapter 9) and is optimized for

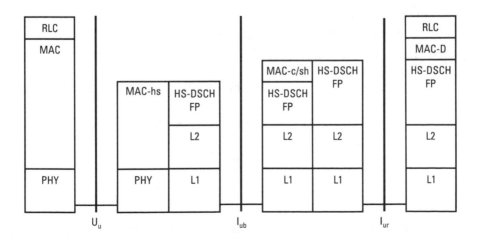

Figure 3.9 HSDPA architecture. (*Source:* [22] © ETSI 2001.)

low- to medium-speed users in urban environments, although also supporting others.

MAC-hs supports a new acknowledgment method known as hybrid ARQ (HARQ) and scheduling specific to HSPDA. Flow control is handled jointly by MAC-c/sh (responsible for normal DSCH) and MAC-hs, and scheduling is based on a priority class passed down from the RRC with separate queues per class.

HARQ is a stop and wait (for acknowledgment) protocol. It uses a new data indicator that is incremented each time that the block transmitted differs from its predecessor; new and old data (i.e., retransmissions) cannot be mixed in the same TTI. Blocks also carry a transmission sequence number per priority class for use in reordering.

The HS-DSCH exists only on the DL and is processed from a single CCTrCH, with only one CCTrCH of HS-DSCH type per UE. HSDPA signaling is sent in-band on the DL in the HS-DSCH, but with out-of-band ACKs or NAKs back from the UE. The UE sends an ACK if the block is correctly received or a NAK if it contains an error. Errored or unacknowledged blocks are retransmitted and take priority over new data. On the DL, the UE identity, the new data indicator, TSN, and priority are included in the MAC header.

At the physical level, the DL channel configuration consists of the PDCH together with up to four separate high-speed shared control channels (HS-SCCH) in combination with the high-speed physical downlink shared channel (HS-PDSCH). The HS-SCCH carries the following:

1. A transport format and resource indicator (TFRI) that shows the transport format, channelization codes, and modulation scheme;
2. The HARQ information.

The TTI is much shorter than the normal 10–80-ms range. There is a single TTI for FDD that is 2 ms, while that for 1.28-Mcps TDD is 5 ms, and that for 3.84-Mcps TDD is 10 ms. The type of error protection coding used is turbo at rate 1/3 with a 24-bit CRC; so a 2-ms block at 2 Mbps would correspond to roughly 1,200 input data bits.

3.6 QoS in WCDMA

This section shows how the above features and procedures control the QoS on the radio link, and what can be achieved with particular reference to the user perception outlined in the first chapter.

The first criterion—that of a high chance of successful access—is determined mainly by capacity planning and admission control (which is discussed in Chapter 6) rather than the RF features above. The time taken to log on to the network and to initiate a call is determined by the cell search and random access procedures. The lack of strict time synchronization in WCDMA means that the three-step search procedure above is used instead of the two-step process common to most other mobile technologies, and this takes slightly longer. There are additional delays where it is necessary to set up a dedicated radio bearer for the user traffic instead of a preexisting common channel.

The next issue is whether calls are liable to be dropped in the middle, especially during handover. WCDMA uses a soft handover procedure in which the UE is in constant communication with several cells and is monitoring their respective signal strengths and qualities. The detailed procedures for performing this should ensure that it is seamless for packet data, but there is a slight possibility of loss of synchronization where onward communication to a CS network is involved.

The QoS profile for the call traffic is determined by the requests of the end applications, the subscription rights of the users, and the available capacity. The ultimate decision on this is taken by the core network and signaled back to the RNC (see Section 8.1.3) in the form of the radio bearer information entity.

The attributes used in the UMTS QoS classes are as follows: maximum bit rate, delivery order, maximum SDU size, SDU format information, SDU error rate, residual bit error rate, delivery of erroneous SDUs, transfer delay, guaranteed bit rate, traffic handling priority, allocation/retention priority, and source statistics descriptor. These are used to select the appropriate radio bearer service between levels 3 and 2. The SDU sizes determine the RLC sizes and level and 1 to 2 transport formats. The following paragraphs show how these factors are represented in terms of parameters for the radio link.

Bit-Rate Throughput is critical for real-time streaming applications and is determined in the absence of retransmissions for a transport channel by the formula

$$\text{Bit rate} = (\text{transport block set size}) \: / \: \text{TTI}$$

The rates achievable depend on the capability of the DPDCH and the choice of spreading factor. The peak data rate on a single DPDCH is 960 Kbps using the minimum spreading factor of 4, but this rate assumes no channel coding; allowing for a coding rate of 1/2 and overheads associated with monitoring for handover, this drops to an effective maximum user data rate of 400 to 500 Kbps, and 300 to 350 at rate 1/3. A further reduction in throughput will result if any additional processing is required (e.g., sequencing and especially retransmissions), so the maximum is most likely to be achieved with RLC TM. Up to six parallel codes, each associated with its own DPDCH, can be used simultaneously (as well as one DPCCH), leading to a maximum possible user data rate of about 2.5 Mbps in conformity with the IMT-2000 target of 2 Mbps.

Delivery Order   This requires the use of sequence numbers by the RLC and hence uses either UM or AM modes.

SDU Error Rate and Bit Error Rate   The BER and BLER on the radio link depend on the power used, code rate, and bit rate. BLER is normally in the range 10^{-1} to 10^{-3} with the power levels that should achieve this under test conditions for various bit rates specified in UMTS [16]. Achievement of better error rates entails use of ARQ retransmissions at the RLC level using RLC-AM. The CRNC is responsible for determining the best combination of power and retransmission count and sets a parameter MaxDAT for the latter after which erroneous SDUs are just discarded.

Delivery of Erroneous SDUs   Distinction between correct and incorrect frames requires the use of a CRC at the physical level.

Transfer Delay   The radio link's contribution to network transfer delay is largely determined by the TTI plus a very small processing element (rather than the radio frame duration) for dedicated channels, plus additional queuing delays (see Chapter 9) for shared channels. The radio frame duration is 10 ms, but the TTI can be 10, 20, 40, or 80 ms, and usually at least 20 ms. All the frames sent in the appropriate TTI are interleaved to reduce errors, so an individual frame cannot be fully processed until the complete set has been received. Turbo-coding entails a few milliseconds of extra delay as compared to convolution owing to internal interleaving between its two stages. For delay-critical traffic such as voice, the 80-ms option is unacceptable in communication with remote networks since the maximum reasonable end-to-end transit time is about 100 ms. Even the 40-ms option is potentially troublesome, so the 20-ms option is used for this. It is therefore important that an application be able to use a suitable TTI (this issue is discussed more fully in the chapters on applications). Transfer delay is increased considerably if

ARQ error correction is required. If N retransmissions are required on average to achieve the required error rate, the transfer delay becomes roughly

$$\text{Transfer delay} = \text{TTI} + N \times (\text{TTI} + Tr)$$

where Tr is a retransmission timer. Several different procedures are possible for this depending on the optional use of a polling bit in RLC AM PDUs with differing timers for retransmission of unacknowledged PDUs [9]. Retransmission of PDUs takes precedence over initial transmissions.

Priority This applies to the common and shared channels. Access service class priority applies in general for the PRACH, while interactive class traffic (see Chapter 9) uses handling priority on shared channels.

Choice of Transport Channel Type

User traffic can be sent over several of the transport channel types in addition to the dedicated traffic channels to ensure efficient use of radio resources; the choice is made by the CRNC. Where a network offers both FDD and TDD, and the UE is capable of both modes, then there is also a choice between these. TDD has two related advantages: the ability to share bandwidth between UL and DL as required, and the ability to support multiple CCTrCH on both DL and UL. These are most useful when the traffic is highly asymmetric and when a user wants support for multiple QoS on the UL. The corresponding applications are receipt of real-time streaming traffic and transmission of multimedia by the UE.

On a dedicated channel the OVSF has to be set on the basis of the highest data rate expected, with the result that they are wasteful if the data has a bursty nature, whereas a shared channel can statistically multiplex bursts from different users, albeit at the expense of some queuing delays (see Section 9.5.1). The dedicated channels also take a relatively long time to set up compared to the common channels, and the latter can sometimes be used to send small amounts of data, such as the Short Message Service (SMS) or e-mail messages. These criteria form the basis of Table 3.5.

The type of channel selected depends in part on the RLC mode, with dedicated channels being used for RLC-AM. CS traffic (e.g., speech and real-time applications) normally make use of dedicated channels. These factors are considered in relation to specific applications and the UMTS QoS classes in Chapter 9.

Table 3.5
Transport Channel Aptitudes

Channel Type	Dedicated	Shared		Common		
Channel Criterion	DCH	DSCH	USCH (TDD only)	FACH	RACH	CPCH
Direction	DL, UL	DL	UL	DL	UL	UL
Soft handover	Yes	No	No	No	No	No
Basis of code usage	Peak rate	User share	User share	Cell	Cell	Cell
Data volume	High/med	High/med	High/med	Low	Low	Med/low
Data-burst suitability	No	Yes	Yes	Yes	Yes	Yes

References

[1] ETSI TS 125 430 "UTRAN I$_{ub}$ Interface; General Aspects and Principles," V.4.1.0 (2001), http://www.etsi.org.

[2] ETSI TS 125 433, "NBAP Specification," V.4.2.0 (2001), http://www.etsi.org.

[3] ETSI TS 125 413, "UTRAN I$_u$ Interface RANAP Signaling," V.4.2.0 (2001), http://www.etsi.org.

[4] ETSI TS 125 301 "Radio Interface Protocol Architecture," V.4.1.0 (2001), http://www.etsi.org.

[5] ETSI TS 125 331 "Radio Resource Control (RRC) Protocol Specification," V.4.2.1 (2001), http://www.etsi.org.

[6] ETSI TS 125 323 "PDCP Protocol Specification," V.4.2.0 (2001), http://www.etsi.org.

[7] ETSI TS 125 324 "BMC Protocol Specification," V.4.0.0 (2001), http://www.etsi.org.

[8] Third Generation Partnership Project, TS 22.146, "Multimedia Broadcast/Multicast Service—Stage 1," http://www.3gpp.org.

[9] ETSI TS 125 322 "RLC Protocol Specification," V.4.2.0 (2001), http://www.etsi.org.

[10] ETSI TS 125 321 "MAC Protocol Specification," V.4.2.0 (2001), http://www.etsi.org.

[11] ETSI TS 125 302 "Services Provided by the Physical Layer," V.4.2.0 (2001), http://www.etsi.org.

[12] ETSI TS 125 211, "Physical Channels and Mapping of Transport Channels onto Physical Channels (FDD)," V.4.2.0 (2001), http://www.etsi.org.

[13] ETSI TS 125 214 "Physical Layer Procedures (FDD)," V.4.2.0 (2001), http://www.etsi.org.

[14] ETSI TS 125 201 "Physical Layer—General Description," V.4.0.0 (2001), http://www.etsi.org.

[15] ETSI TS 125 212 "Multiplexing and Channel Coding (FDD)," V.4.2.0 (2001), http:// www.etsi.org.

[16] ETSI TS 125 101, "WCDMA-FDD UE Radio Transmission and Reception," V.4.2.0 (2001), http://www.etsi.org.

[17] ETSI TS 125 215 "Physical Layer Measurements (FDD)," V.4.2.0 (2001), http://www.etsi.org.

[18] ETSI TS 125 304 "UE Procedures in Idle Mode and Procedures for Cell Reselection in Connected Mode," V.4.2.0 (2001), http://www.etsi.org.

[19] ETSI TS 125 214, "Physical Layer Procedures (FDD)," V.4.2.0 (2001), http://www.etsi.org.

[20] ETSI TS 125 133 "Requirements for Support of Radio Resource Management (FDD)," V.4.2.0 (2001), http://www.etsi.org.

[21] ETSI TS 125 221 "Physical Channels and Mapping of Transport Channels (TDD)," V.4.2.0 (2001), http://www.etsi.org.

[22] Third Generation Partnership Project, TS 25.308 "UTRA High Speed Downlink Packet Access (HSDPA)," Rel 5.2, http://www.3gpp.org.

4

cdma2000

4.1 General Principles

Cdma2000 is a direct sequence spread spectrum code division multiple access system that provides a natural evolutionary path from cdmaOne, as defined in IS-95A/B, to full 3G functionality. The IS-95 standard was originally defined by Qualcomm in the United States to reuse the 800-MHz cellular band used by AMPS with CDMA's higher performance. Effectively a number of 30-kHz carriers of AMPS were replaced by a single IS-95 carrier of 1.25 MHz bandwidth, giving roughly one CDMA channel per 40 AMPS channels. Channel bit rates for this service were chosen to match the bit rate requirements of the CELP voice codec used by mobiles; that is, 9.6 Kbps and submultiples thereof. A parallel development, once called CDMA-PCS, was also carried out for mobile operators using the 1,900-MHz PCS band in the United States. This version uses a higher-quality voice codec that needs higher channel bit rates, with 14.4 Kbps and submultiples chosen. This distinction is carried forward in the cdma2000 specifications as the basis for rates sets 1 and 2, while the distinction between types of mobile forms the start of a series of radio configurations. Both of these topics are featured in the description of cdma2000 that is given in the following sections. Cdma2000 is evolving through a series of stages, and this book is based largely on Release A functionality, while a guide to Release B is given by [1].

The characteristic feature of cdma2000 is its use of multiple narrowband carriers instead of the single wideband carrier of WCDMA. Further distinctions between cdma2000 and the earlier cdmaOne include the use of a more efficient QPSK modulation (which is a prime distinction between cdmaOne and cdma2000 1X), support for variable bit rates, and better power control. Cdma2000 is evolving through a sequence of phases that use successively larger numbers of these narrowband carriers; thus, cdma2000 1X uses a single 1.25-MHz carrier, while cdma2000 3X uses three such 1.25-MHz carriers via distinct base station antennae to provide 3.75 MHz, with probable 6X, 9X, and 12X versions to follow in due course. An advantageous side effect of this is the provision of both frequency and antenna diversity, and consequently extra resilience.

Another major difference between cdma2000 and WCDMA is that the former requires strict synchronization of base stations based on GPS clocks. For macrocells, such as are likely in rural areas, this does not constitute a limitation, but for urban areas with microcells of a few street blocks there may be some difficulty in finding base sites with sufficient line-of-sight access to GPS.

A major aspect of the philosophy behind cdma2000 is to make as much use of existing standards and technologies as possible; and in keeping with this, it uses mobility IP as the basis for transmission through the core network. The physical architecture of a cdma2000 network is shown in Figure 4.1.

This chapter is concerned with the radio access section. The interaction of the access with the core section is covered in Chapter 8. The cdma2000 radio access network has a three-layer structure as indicated in Figure 4.2,

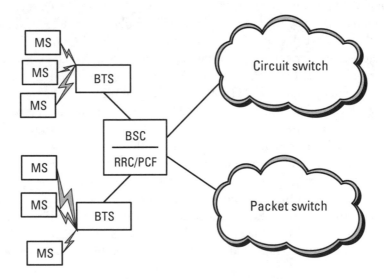

Figure 4.1 Cdma2000 network architecture.

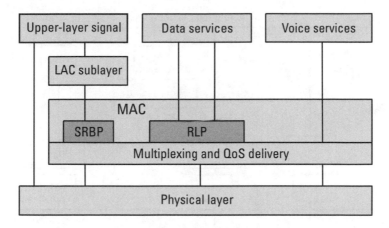

Figure 4.2 Protocol architecture. (*After:* [2].)

but both the MAC and the top link access control (LAC) belong to OSI layer 2. The functionality of each layer is described in the following sections.

4.2 Layer 3 Signaling and LAC

OSI level 3 to 7 functions are contained in a large range of signaling messages between the base station and mobile or vice versa. The base station issues signaling messages on several channels—notably broadcast, paging, and forward common/dedicated signaling (see Section 4.3.2)—to provide configuration and control information. Messages that are sent by the mobile that are not responses to messages from the base are called autonomous messages. These messages are passed to layer 2 via the LAC sublayer.

The structure of the LAC layer is shown in Figure 4.3 for the forward logical channels; that for the reverse direction lacks the sync and broadcast channels. The specification of LAC operation is split into two parts, one handling signaling messages [3] and the other dealing with call control [4].

The top sublayer for the reverse channel signaling is for authentication and has little impact on QoS issues. This is used to identify the mobile at various stages in a call, notably registration and termination. It is based on three main components: a 32-bit electronic serial number configured into the mobile at manufacture, a 64-bit secret data authorization field, and a 32-bit random challenge field, plus an additional authorization field that is usually set to the 24-bit IMSI-S1 (the last seven digits of the 10-digit International Mobile Station Identifier).

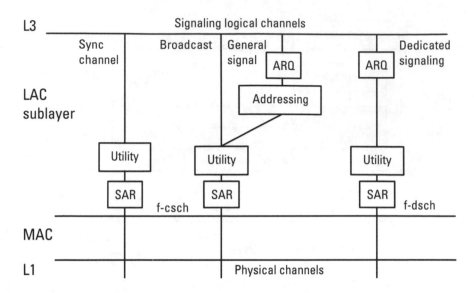

Figure 4.3 UL and LAC signaling. (*After:* [3].)

The next sublayer (both forward and reverse) is the automatic repeat request (ARQ), which handles retransmissions. Messages can be sent in either of two modes, assured or nonassured, of which the former makes use of acknowledgments and retransmissions while the latter does not. Voice and data traffic for established calls bypasses the LAC and goes directly from OSI level 3 to the MAC, while all signaling goes via the LAC, then onward by one of a number of logical signaling channels to the MAC. There are three overall types of message that the ARQ can receive from higher layers:

1. A request-type selection/distribution unit (SDU) to be transmitted and its subtype (e.g., registration). In such cases, the ARQ sublayer does not include an acknowledgment within the PDU to be transmitted.

2. A response-type SDU to be transmitted, its subtype (e.g., registration), identifying information to allow the pairing of the message to be transmitted with the received message to which it is replying, and possible information on how the ACK_TYPE should be set in the reply PDU. In such cases, the ARQ sublayer on the transmitter side includes an acknowledgment in the PDU if the ARQ sublayer on the receiver side indicates that an acknowledgment is required.

3. An indication that no response SDU is outstanding. In such cases, the ARQ sublayer on the transmitter side may generate an acknowledgment-only PDU if the ARQ sublayer on the receiver side indicates that an acknowledgment is required.

For most of the signaling messages an acknowledgement is required, and the procedure varies slightly according to the signaling channel. The signaling channels are as follows:

- Forward or reverse common signaling channels (F/R-CSCH);
- Forward or reverse dedicated signaling channels (F/R-DSCH).

Forward refers to the downlink from the base station to the mobile, and reverse refers to the uplink.

The next main sublayer is that of addressing, which is vital for identification but of little significance for performance. The addressing sublayer inserts the addresses of mobile stations in common channel signaling. The utility sublayer inserts a message-identity, messageID type, and messageID length in the LAC PDU based on the message tag in the MCSB for L3 to L2 or conversely for L2 to L3. MessageID types can be ESNs, international mobile subscriber identifiers (IMSIs), temporary mobile subscriber identifiers (TMSIs), ESNs and IMSIs, or ESNs and IMSI-S (a 10-digit derivative of the IMSI). In addition, some other field, such as IMSI class, may be added. This layer also reports pilot signal measurements to the RRC. There are more than a dozen message types (e.g., registration), but none are directly related to QoS.

Segmentation and reassembly (SAR) is of more significance for this. There are essentially two types of PDU sent by the LAC: regular and mini-PDUs. The mini-PDUs are never segmented and have a fixed length of 48 bits for transmission in 5-ms radio frames at the physical level. Regular PDUs have to be segmented in order to fit into the normal 20-ms frames at the physical level. This is done by splitting the LAC PDU into fragments of the maximum permissible size for the physical channel, setting a start of message (SOM) indicator, a message length field, and adding a 30-bit CRC. In general, fragments have to be padded to achieve a length of $8k + 2$ bits (excluding CRC), where k is an integral number and the maximum length is 842 bits or less, depending on physical channel parameters. The SAR is also responsible for removing the message control and status block (MCSB) sent from L3 to L2 or inserting it in the opposite direction. The MCSB contains the QoS parameters and is described in Section 4.5.

In general, the operation of these sublayers varies according to the type of signaling channel. A very brief outline of each is given below.

Common Signaling Channels

Acknowledgments on the common signaling channels are controlled by the ARQ using the fields shown in Table 4.1.

MSQ_ACK is used by the receiver to recognize duplicate messages via equivalent values within a timer T4. An access attempt for a message consists of an initial transmission and a configurable number of retransmissions, each transmission being called an access probe. If the probe is not acknowledged within another timer, configurable for values between 160 and 1,360 ms, it is retransmitted.

Dedicated Signaling Channels

ARQ fields on the dedicated channels are fairly similar to those on the common channels, as shown in Table 4.2.

The mini-PDUs are more time critical than the regular PDUs and so they use different timers and retransmission counts, usually quicker and more numerous.

Table 4.1
Acknowledgment Fields on R-CSCH

Acronym	Size (bits)	Significance
ACK_SEQ	3	Value of MSQ_SEQ received on f-csch being acked on R-CSCH
MSQ_SEQ	3	Sequence number of PDU to send on R-CSCH
ACK_REQ	1	Indicates if PDU on R-CSCH will need acknowledgment
VALID_ACK	1	Set if PDU on R-CSCH is acknowledging a PDU on f-csch
ACK_TYPE	3	Address type or page class of message acknowledged

Table 4.2
Acknowledgment Fields on R-DSCH

Acronym	Size (bits)	Significance
ACK_SEQ	2 or 3	Value of MSQ_SEQ received on f-dsch being acked on R-DSCH
MSQ_SEQ	2 or 3	Sequence number of PDU to send on R-DSCH; 2 is for mini-PDU
ACK_REQ	1	Indicates if PDU on R-DSCH will need acknowledgment

A dedicated channel does not need additional addressing fields to identify the mobile.

The most important message types for QoS are those associated with obtaining extra bandwidth for a call, such as the supplemental channel request messages/supplemental channel assignment messages (SCRMs/SCAMs) and their extended counterparts. These request or acknowledge assignment of an additional traffic channel for the same mobile, allowing it to increase its traffic throughput. The number of supplemental channels available depends on the radio configuration of the mobile as described in the physical layer section of this chapter (Section 4.4). The extended versions support more QoS parameters.

There are also corresponding minimessages, the SCRMM and SCAMM. The short minimessages achieve a more rapid response than the regular versions.

4.3 Layer 2 MAC

4.3.1 General

Data and voice traffic from the network bypasses the LAC and is processed directly by the MAC layer via the Radio Link Protocol (RLP) and multiplexing sublayers, while signaling is passed from the LAC to the connectionless Signaling Radio Burst Protocol (SRBP) sublayer (see Figure 4.2), which uses common logical channels. The MAC layer can provide both a reasonably reliable best efforts service based on the RLP and a more strictly controlled QoS approach using multiplexing and prioritization schemes. Data traffic usually uses the F/R-DTCH channel and RLP, while CS traffic is sent directly via the F/R-DTCH. RLP is connection oriented and provides reliability through negative acknowledgments, but it does not support QoS.

The MAC communicates with the LAC through responses to LAC signaling requests. General signaling is sent on the common channels, while signaling for specific MS goes on the dedicated signaling channel. These requests are of the form shown below to indicate the parameters on which they depend. There are two request types and two responses.

MAC-SDU Ready Request sent by the LAC carries the following parameters: channel type, SDU size (bits), sequence number, and scheduling hint. The response from the MAC is a MAC-Availability Indication with the following parameters: channel type, maximum size (bits of SDU on L1), and system time. Alternatively, the MAC may return a MAC-Access-Failure, giving the reason for the failure. On receipt of MAC-Availability, the LAC sends a MAC-Data Request with the following parameters: channel type, data (SDU or fragment to transmit), size (of data in bits). When the MAC receives data from the physical layer, it sends data to the LAC via a MAC-Data Indication with the following parameters: channel-ID, channel type, data (SDU received), size (of data in bits), system time.

One of the most basic parameters is the channel type, for which the possible options are given in the next section.

4.3.2 Channel Types

Channel types consist of logical channels for higher layers and physical channels for level 1. The logical channels can be further subdivided into signaling channels and data channels, with the signaling channels as described in the previous section.

The channels requested in the higher message primitives above are the appropriate physical channels:

- Forward synchronization (timing and framing to mobile station) (F-SYNC);
- Forward broadcast common channel (F-BCCH);
- Forward common access channel (F-CACH);
- Reverse enhanced access channel (R-EACH);
- Forward common power control channel (F-CPCCH);
- Forward/reverse common control channel (F/R-CCCH);
- Reverse assignment channel (R-ACH);
- Forward paging channel (F-PCH).

In addition, the dedicated logical channels make use of the following physical channels:

- Forward/reverse dedicated control channel (F/R-DCCH);
- Forward/reverse fundamental channel (F/R-FCH);
- Forward/reverse supplemental channel (F/R-SCH);
- Forward/reverse supplemental code channel (F/R-SCCH).

The fundamental channel is a basic traffic channel at 9.6 Kbps that is automatically provided for a call. Supplemental channels are additional channels at 9.6 or 14.4 Kbps, up to seven of which can be offered. Supplemental code channels (SCCHs) are defined as in IS-95A/B. In cdma2000 there are up to two high-speed SCHs for use with the later radio configurations instead of the SCCH.

In the message primitives above, the channel-ID is the specific instance of the channel type involved. The System_Time in the MAC-Availability is an estimate of when the physical layer will transmit the first bit of the data, and in the MAC-Data Indication denotes when the physical layer received the first data bit. The Scheduling-Hint from the LAC gives information on priority to apply to the SDU in the multiplexing sublayer of the MAC.

4.3.3 The Multiplexing Sublayer

This layer handles the multiplexing/demultiplexing roles between the physical layer and the top of the MAC, taking into account QoS considerations for the various traffic types. This is shown in overview in Figure 4.4.

This figure shows signaling traffic being multiplexed with some of the data traffic for a single user onto the FCH or DCCH using an appropriate multiplexing option, while another two blocks of the same user's data are first made into MAC PDUs of a suitable multiplexing option, then further combined for transmission over an SCH. In the figure, H and M denote MAC headers of different types, while C denotes a CRC. The multiplexing option to be used is included in the MCSB.

There are several different ways to perform this multiplexing, which depend on the historical evolution of the standards and the increasing range

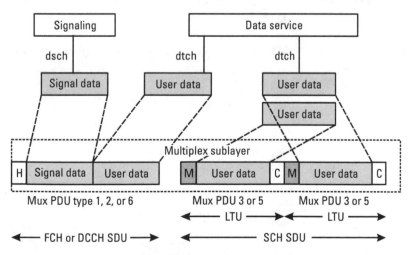

Figure 4.4 Multiplexing example. (*Source:* [2] © 3GPP2.)

of different traffic types that have to be accommodated. A key factor in this is the radio configuration (RC) of the mobile. This determines transmission rates, modulation characteristics, and spreading rates that the mobile can support. The original IS-95 mobiles have RC 1, while those for the 14.4-Kbps services have RC 2. A number of additional configurations have been defined for later versions of cdma2000, and these can be treated according to their classes as follows:

1. RC class 1 use RC 1 or RC 2 on both the forward and reverse traffic channels;

2. RC class 2 use RC 3 or RC 4 on the reverse traffic channels, and RC3, RC 4, or RC 5 on the forward channels;

3. RC class 3 use RC 5 or RC 6 on the reverse channels, and RC 6, RC 7, RC 8, or RC 9 on the forward channels.

The sublayer has two modes of operation, A and B, depending on the physical capability of the mobile as indicated by its RC class. RC class 1 uses mode A, while higher values (for cdma2000 functionality) use mode B.

There are six different multiplexing options as indicated by the MuxPDU type values 1 to 6. The types of MuxPDU that can be supported on the various physical channels are given in Table 4.3.

Multiplexing also depends on the selection of one of two rate sets. Rate set 1 consists of 9.6, 7.2, 4.8, 2.7/2.4, 1.5/1.2, and all integral multiples of 9.6 Kbps. This set evolved from the original IS-95 for the 800-KHz band. Rate set 2 consists of 14.4, 7.2, 3.6, 1.8, and all integral multiples of 14.4 Kbps, as required for CDMA-PCS.

The precise manner of multiplexing is determined by the physical channel type and by a multiplexing option appropriate to the rate. Rates in cdma2000 are much less flexible than in WCDMA—where any rate can be

Table 4.3
Multiplexing Options

Channel Type	MuxPDU Types
FCH	1, 2, 4, 6
DCCH	1, 2, 4, 6
SCCH	1, 2
SCH	1, 2, 3, 5

Source: [2].

achieved by power control—and are determined through the application of specific rate compatible punctured convolutional (RCPC) codes to the input PDUs. Many tables define the correct combinations. These tables show how different traffic types, such as signaling and different categories of user-plane data, are combined using fields of specified position and size. A table is shown below for the original simple case of a 9.6-Kbps channel and MuxPDU type 1. Multiplexed traffic in this case goes into the 172-bit information field of a 20-ms duration 192-bit radio frame along with a 12-bit CRC and 8-bit frame tail delimiter for the encoder. Table 4.4 shows the possible multiplexing options within the 172-bit information frame at 9,600 bps.

TM indicates the traffic-mixing ratio of primary to signaling/secondary. The phrase *dim and burst* is sometimes used to indicate options where the frame contains both primary and signaling/secondary traffic; the phrase *blank and burst* is used for frames with signaling/secondary only.

For rate set 2 frames at 14.4 Kbps, there is an analogous set of MuxPDU type 1 options, but also a MuxPDU type 2 that differs in the way of distinguishing between secondary and signaling traffic types. Here the 20-ms frames consist of 288 bits, of which the information field contributes 267. A 4-bit frame mode (FM) field that shows the ratio between separate primary, signaling, and secondary fields replaces the TM field. SCCH channels only support these two multiplexing sets of options.

Table 4.4
MuxPDU Type 1 Fields and Bit Sizes

Option	MM	TT	TM	Primary Traffic	Signaling	Secondary Traffic
1	0	n/a	n/a	171	0	0
2	1	0	0	80	88	0
3	1	0	1	40	128	0
4	1	0	2	16	152	0
5	1	0	3	0	168	0
6	1	1	0	80	0	88
7	1	1	1	40	0	128
8	1	1	2	16	0	152
9	1	1	3	0	0	168

Notes: The term *primary* refers to speech traffic, and the signaling field can contain either signaling or secondary data, such as FAX. MM denotes mixed-mode flag, with 0 indicating unmixed and 1 indicating mixed. TT is a 1-bit secondary traffic-type indication. *Source:* [2].

As the cdma2000 standards have evolved for cdma2000 1X and 3X, additional MuxPDU types have been defined to meet still more complex needs, particularly larger block sizes and higher transmission rates on the supplemental channels. The information block sizes in bits supported for rate sets 1 and 2 are shown in Table 4.5. Together with the transmission times of 20, 40, or 80 ms, these block sizes determine the rates of data transmission on the supplemental channel.

The multiplex sublayer converts data bits from signaling into data blocks, then multiplexes one or more data blocks into a MuxPDU, as described above, and then combines one or more MuxPDUs into a physical level SDU according to channel type. For the FCH and DCCH there is only a single MuxPDU per SDU, and the same is true for SCCH in the case of mode A. Multiple MuxPDUs can go in one SDU for an SCH in mode B.

Mode A Operation

If an FCH has been allocated, then an FCH SDU is assembled every 20 ms, where for rate set 1 the data is made into a MuxPDU of type 1, while for rate set 2 a MuxPDU of type 2 is used.

For each SCCH that has been assigned, a 20-ms SCCH SDU is assembled every 20 ms with the same types as for the FCH.

Mode B Operation

The operation of mode B is more complex: first, an FCH SDU is only assembled at the start of 20-ms periods, and second, to make a 20-ms FCH SDU all data blocks from the signaling channel and logical channels assigned

Table 4.5
Information Block Sizes

Configurations	RC 3, 5 on R-SCH RC 3, 4, 6, 7 on F-SCH	RC 4, 6 on R-SCH RC 5, 8, 9 on F-SCH
Block sizes (SDU)	172	267
	360	552
	744	1,128
	1,512	2,280
	3,048	4,584
	6,120	9,192
	12,264	20,712

Source: [4].

to the FCH must be combined. MuxPDU types 1 or 6 are used for rate set 1, and MuxPDU types 2 or 6 are used for rate set 2. It can also handle 5-ms FCH SDUs via MuxPDU type 4.

Where a DCCH has been assigned, mode B will also generate 20-ms DCCH SDUs at the start of 20-ms periods using MuxPDU types 1 or 6 for rate set 1 and MuxPDU types 2 or 6 for rate set 2.

For each SCH that has been assigned, the multiplex sublayer will assemble an SCH SDU every 20, 40, or 80 ms, according to the SCH frame length configured. This is done by forming a MuxPDU from a single data block for any SCH assigned logical channel in order of channel priority, then combining them to form the relevant SDU size or until all available MuxPDUs have been combined. The type of MuxPDU used depends on the multiplexing option.

Logical Transmission Units

In the cases of MuxPDU types 3 and 5, a number of MuxPDUs are combined into a single logical transmission unit (LTU) containing a 16-bit CRC error correction field. The size of these LTUs is smaller than the physical SDU size on which it depends. For type 3 the LTU size is 368 bits for 744, 1,512, or 3,048 bits and 560 bits for 1,128, 2,280, or 4,584 bits, with a more complex set of options for type 5. The multiplex sublayer determines the number (defaults 0, 2, 4, or 8) of LTUs per SDU information block (with sizes as per Table 4.5 for the SCH) in mode B.

The MAC layer assigns headers to the MuxPDUs according to their type. Those for types 1 and 2 are the same as those in IS-95B, while type 3 has a 6-bit header including a 3-bit service-instance identifier. The 5-ms type 4 units have no header, while those for types 5 and 6 are of variable size and include information on length for type 5 and other options for type 6.

Interface Messages to Physical Layer

The MAC communicates with the physical layer via transmit requests or receive indications for the various physical channels in operation.

4.4 Layer 1 Physical Level

4.4.1 Carriers

The characteristic feature of cdma2000 is its use of multiple carriers in general. The first version, cdma2000 1X, only uses a single carrier of 1.228 Mcps for both forward and reverse directions. The next version is cdma2000 3X, which uses three carriers at 1.228 Mcps in the forward direction and one at 3.684 Mcps in the reverse, while later versions will use larger multiples of 1.228 Mcps in the same way.

The radio configuration of a mobile is related to its ability to support these options. Configurations 1 and 2 refer to the earlier IS-95 kit supporting rate sets 1 and 2, respectively, while configurations 3 and 4 correspond to cdma2000 1X.

4.4.2 Frequency Bands

Another of the main characteristics of cdma2000 is the support for a wide range of frequencies. The approximate frequency bands [1] and their other usages are shown in Table 4.6.

Where the mobile and base station frequency bands overlap in this table, the bands are split into subbands within which there is no overlap. The base station frequency is always higher than the mobile except for bands 3 and 7 where the reverse is true. The number of channels that can be supported varies according to the frequency band, with the unallocated band 8 offering the most. Channel numbering is based on the old 30-kHz AMPS scheme, so the nominal 1,450 channels for band 8 translates into about 35 1.25-MHz channels. The precise number varies according to the spreading rate to be used. The cdma2000 standards give recommendations for the way

Table 4.6
Frequency Bands

Band	Name	Mobile Frequency (MHz)	Base Frequency Number (MHz)
0	Cellular	824–850	869–894
1	North America PCS	1,850–1,910	1,930–1,990
2	TACS	872–915	917–960
3	JTACS	887–925	832–870
4	Korean PCS	1,750–1,780	1,840–1,870
5	NMT-450	411–484	421–494
6	IMT-2000	1,920–1,980	2,110–2,170
7	North America 700 Cellular	776–794	746–764
8	1,800 MHz	1,710–1,785	1,805–1,880
9	900 MHz	880–915	925–960

in which the individual channel numbers are related to subband frequencies. It may be noted that the North American PCS systems in band 1 overlap with the bottom of the IMT-2000 recommended frequency for UMTS in band 6, providing part of the motivation for not following the IMT-2000 plans fully. In order to comply with cdma2000, a mobile must support at least one of the above bands, but it does not have to support them all, so operators of international networks can potentially have a problem.

4.4.3 Timing

Times are based on a system time obtained by a base station from GPS. Mobiles derive their system times by applying their propagation delay to the base station's time. Synchronization of the base stations to a common GPS time standard means that mobiles are able to identify different base stations by the phase of their signals. WCDMA uses 512 scrambling codes for this purpose, and cdma2000 makes use of 512 possible phases.

A frame is the basic timing interval in the system. For the forward and reverse supplemental code channels—and also the access and paging channels—it is 20 ms. For the forward supplemental channel and the reverse supplemental channel, a frame is 20-, 40-, or 80-ms long. For the enhanced access channel, the forward common control channel, and the reverse common control channel, a frame is 5-, 10-, or 20-ms long. For the forward fundamental channel, forward dedicated control channel, reverse fundamental channel, and reverse dedicated control channel, a frame is 5- or 20-ms long. For the common assignment channel, a frame is 5-ms long.

Frame Quality Indicator

This is the CRC check applied to 9.6- and 4.8-Kbps traffic channel frames of RC 1, all forward traffic channel frames for RC 2 through 9, all reverse traffic channel frames for RC 2 through 6, the broadcast control channel, common assignment channel, enhanced access channel, and the reverse common control channel.

4.4.4 Power Control

Power control in cdma2000 differs from that in WCDMA in several respects, most notably in the former's use of fast power control on the UL alone, and in the manner that it varies with bit rate. For the fixed rate channels in cdmaOne/cdma2000, the energy per symbol is reduced according to the number of repetitions so that the energy per data bit remains constant.

There are three independent means of adjusting the power of a mobile station: open loop estimation by the mobile, closed loop by the mobile and base together, and for RC 3 to 6 also a code channel attribute adjustment by the two. Detailed procedures and values are defined for each of these. The closed loop method depends on power control bits received on the relevant channel, with fast power control operating at 800 Hz. Open loop control is used less extensively in cdma2000 than in IS-95, but may still supplement the fast power control to cater for sudden environmental effects, such as going around a corner. Outer loop control is applied on a per-channel basis, whereby the mobile adjusts its power within a permissible range to achieve the target error rate.

4.4.5   Error Correction

The first part of error correction is the addition of the CRC frame quality indicator to a radio frame. The size of this depends on the frame size, the channel, and the data rate, and is typically, but not always, 12 bits for frames under about 360 bits and 16 bits for larger ones. The CRC is omitted on the lowest rate channels (2.4 Kbps or less for rate set 1) because these low rates are used just for comfort noise during silent spells in voice calls.

FEC is applied to the radio frames after addition of the CRC using either convolution or turbo codes according to circumstances. The convolution codes are most often of rate 1/4 and used on all frames up to 360 bits, with the option of turbocoding for larger frames on some channels, but they also depend on the radio configuration and data rate.

The symbol rate generated by the encoder is equal to the data rate divided by the encoder rate, and this is typically less than the symbol rate for the channel. The rate is boosted to the channel symbol rate where necessary by adding an integral number N of repetitions of each symbol so that

$$\text{Channel symbol rate} = \text{encoder symbol rate} \times (N + 1)$$

The target frame error rate as indicated by the quality indicator is 0.05 for most circumstances.

4.4.6   Data Rates

The data rates for the channels depend on the version 1X or 3X, the capabilities of the radio configuration, the symbol rates, and the repetition rates [5]. Radio configurations 1 and 2 can only support fixed rates, but flexible rates are possible with configurations 3 to 6. The flexible rates are associated with frame formats where the numbers of information bits, reserved bits, and frame quality indication bits are configurable. The SCH supports variable bit rates where the rate can be changed on a frame-by-frame basis from the allowed values in Tables 4.7 and 4.8.

In addition to the above rates, it is almost certain that an additional set of rates will be added based on the 12.2-Kbps rate of the AMR codec recommended in the UMTS standards. Analogous tables are available for forward channels [5], and include RC7, 8, and 9.

When fixed data rates are used, there are specific rules relating the number of bits in a frame to the data rate, channel type, and radio configuration. Flexible data rates are achieved where appropriate by allowing a variable number of bits in the frame.

Cdma2000 does not use a data rate indication analogous to WCDMA's transport format indication; rather, it deduces the speed at low rates by performing the Viterbi decoding for multiple possible rates and selecting the one that shows the minimum number of errors. At higher rates the channel rate is given in a service connect message for the traffic channel or extended channel assignment messages (see the following section).

4.5   cdma2000 QoS

4.5.1   Basic Standard

Cdma2000, as defined up to version 3X Release A, has less explicit support for QoS distinctions than does WCDMA. As cdma2000 is backwards compatible with cdmaOne, the quality available depends on the generations of equipment involved; for example, IS-95 B dedicated channels are limited to one fixed rate FCH and up to seven fixed rate SCH in either direction, giving a maximum possible data rate of 76.8 Kbps, whereas variable rates up to 614 Kbps for 1X and 2 Mbps for 3X are possible. From a data user's perspective, the most important QoS factor is likely to be the data rate (Table 4.6), and this depends heavily on the mobile's radio configuration. Because of the large number of possibilities, the message transmission must be preceded by an exchange of service request and response messages between the mobile and base stations to establish the radio configuration, multiplexing options, and frequency bands that can be supported, with the final selection to be used being passed back from the base to the mobile in the service connect message. For radio configuration classes 2 and 3 this will indicate the actual transmission rate to be used on the FCH or SCH. Frequently the mobile is

Table 4.7
Data Rates for Spreading Rate 1 (cdma2000 1X)

Channel Type		Data Rates (bps)
Access channel		4,800
Enhanced access channel	Header	9,600
	Data	38,400 (5-, 10-, or 20-ms frames),
		19,200 (10- or 20-ms frames), or
		9,600 (20-ms frames)
Reverse common control channel		38,400 (5-, 10-, or 20-ms frames),
		19,200 (10- or 20-ms frames), or
		9,600 (20-ms frames)
Reverse dedicated control channel	RC 3	9,600
	RC 4	14,400 (20-ms frames) or
		9,600 (5-ms frames)
Reverse fundamental channel	RC 1	9,600, 4,800, 2,400, or 1,200
	RC 2	14,400, 7,200, 3,600, or 1,800
	RC 3	9,600, 4,800, 2,700, or 1,500 (20-ms)
		9,600 (5-ms frames)
	RC 4	14,400, 7,200, 3,600, or 1,800 (20ms) or
		9,600 (5-ms frames)
Reverse supplemental code channel	RC 1	9,600
	RC 2	14,400
Reverse supplemental channel	RC 3	307,200, 153,600, 76,800, 38,400, 19,200, 9,600, 4,800, 2,700, or 1,500 (20-ms frames)
		153,600, 76,800, 38,400, 19,200, 9,600, 4,800, 2,400, or 1,350 (40-ms frames)
		76,800, 38,400, 19,200, 9,600, 4,800, 2,400, or 1,200 (80-ms frames)
Reverse supplemental channel	RC 4	230,400, 115,200, 57,600, 28,800, 14,400, 7,200, 3,600, or 1,800 (20-ms frames)
		115,200, 57,600, 28,800, 14,400, 7,200, 3,600, or 1,800 (40-ms frames)
		57,600, 28,800, 14,400, 7,200, 3,600, or 1,800 (80-ms frames)

Source: [5].

Table 4.8
Data Rates for Spreading Rate 3 (cdma2000 3X)

Channel Type		Data Rates (bps)
Enhanced access channel	Header	9,600
	Data	38,400 (5-, 10-, or 20-ms frames), 19,200 (10- or 20-ms frames), or 9,600 (20-ms frames)
Reverse common control channel		38,400 (5-, 10-, or 20-ms frames), 19,200 (10- or 20-ms frames), or 9,600 (20-ms frames)
Reverse dedicated control channel	RC 5	9,600
	RC 6	14,400 (20-ms frames) or 9,600 (5-ms frames)
Reverse fundamental channel	RC 5	9,600, 4,800, 2,700, or 1,500 (20 ms) or 9,600 (5-ms frames)
	RC 6	14,400, 7,200, 3,600, or 1,800 (20 ms) or 9,600 (5-ms frames)
Reverse supplemental channel	RC 5	614,400, 307,200, 153,600, 76,800, 38,400, 19,200, 9,600, 4,800, 2,700, or 1,500 (20-ms frames); 307,200, 153,600, 76,800, 38,400, 19,200, 9,600, 4,800, 2,400, or 1,350 (40-ms frames); 153,600, 76,800, 38,400, 19,200, 9,600, 4,800, 2,400, or 1,200 (80 ms-frames)
	RC 6	1,036,800, 460,800, 230,400, 115,200, 57,600, 28,800, 14,400, 7,200, 3,600, or 1,800 (20 ms); 518,400, 230,400, 115,200, 57,600, 28,800, 14,400, 7,200, 3,600, or 1,800 (40-ms frames); 259,200, 115,200, 57,600, 28800, 14,400, 7,200, 3,600, or 1,800 (80-ms frames)

Source: [5].

instructed to start with a default configuration based on multiplexing option 1 prior to receipt of the service connect message.

The rates and frame error rates (in extended cases) are included in channel assignment messages (CAM), extended CAM (ECAM), SCAM, and extended SCAM (ESCAM). CAM and ECAM apply to the initial channels, such as the fundamental channel, while SCAM and ESCAM apply to the extra channels acquired to support high data rates. ECAM and ESCAM are only supported by mobiles capable of configurations additional to RC 1 and 2.

The BTS can assign a mobile up to seven supplemental code channels for RC 1 or 2, but only a maximum of two supplemental channels (at much higher speed) for the later configurations.

Rates for the SCH are notified in the ESCAMs and are quoted with their rate fields in Table 4.9 [4]. They also correspond to specific combinations of multiplex option and radio configuration, which determine rates in general.

RC 1 and 2 only support fixed rate supplemental code channels at 9.6 and 14.4 Kbps, respectively. The LAC also specifies the target frame quality to be used by forward link power control (FPC) in the ESCAM. This is based on a 5-bit field with values as shown in Table 4.10.

The message includes a start time and duration of from 1 to 256 units of 20 ms or infinite (i.e., until another control message). This allows data bursts to be sent at variable rates over the supplemental channels by subsequent assignments with different parameters.

In general, information about QoS can be conveyed in a variable length QoS information record [4] in messages on the signaling channels. This contains subscription data, whether the service is assured or nonassured, and the service option, which contains the traffic channel type, rate, and reliability needs in a service option connection record. Some mobiles are permitted to request QoS parameters in the origination message at the start of call, and a

Table 4.9
Data Rates for Forward/Reverse SCH

FOR/REV_SCH_RATE	Rate in Kbps	
	(a)	(b)
0000	9.6	14.4
0001	19.2	28.8
0010	38.4	57.6
0011	76.8	115.2
0100	153.6	230.4
0101	307.2	259.2
0110	614.4	460.8
0111	Reserved	518.4
1000	Reserved	1,036.8

Note: All other values are reserved. (a) and (b) represent radio configurations as follows: (a) R-RC 3, 5 and F-RC 3, 4, 6, 7; and (b) R-RC 4, 6 and F-RC 5, 8, 9.
Source: [4].

Table 4.10
Target FER Coding

FER_SCH Values	Frame Error Rates (%)
00000	0.2
00001–10100	0.5–10 (in units of 0.5%)
10101–11001	11–15 (in units of 1%)
11010–11110	18–30 (in units of 3%)

Source: [4].

1-bit MOB_QOS field in the extended system parameter message sent by the base station on the paging channel indicates this permission. This flag is also used in autonomous messages, such as the origination message on the R-CSCH that has a 1-bit field to indicate whether QoS parameters are included, and if so, then the length of the field and the QoS information record. The ECAM is sent by the BTS in response to an origination message or paging response from the MS and may contain QoS parameters that differ from those requested.

The stage at which QoS is implemented on the radio link is the multiplexing and QoS sublayer of the MAC [2]. The information used to decide the QoS is included in the MCSB passed to the MAC by layer 3 along with the SDU in the MAC.Data Ready primitive. This states the channel type to use, whether assured or nonassured mode is to be used, and a relative priority in the scheduling hint to use for multiplexing. When assured mode is used, the SDU is transmitted multiple times until an acknowledgment is received from the remote LAC; for nonassured mode the SDU is also repeated several times to allow a good chance of receipt with duplicate detection to remove surplus copies.

In the case of PS traffic, the RADIUS server (see Section 8.3 and [6]) holds a DiffServ (see Section 7.4.5) class option for the MS that determines forwarding priority on shared PS networks and on the shared forward channels.

4.5.2 High Data-Rate Enhancements

Two sets of enhancements to cdma2000 1X have been proposed for the support of high-speed data services without using multiple carriers: 1xEV-DO [7] and 1xEV-DV [8]. The former uses its own protocol stack, which is shown in Figure 4.5.

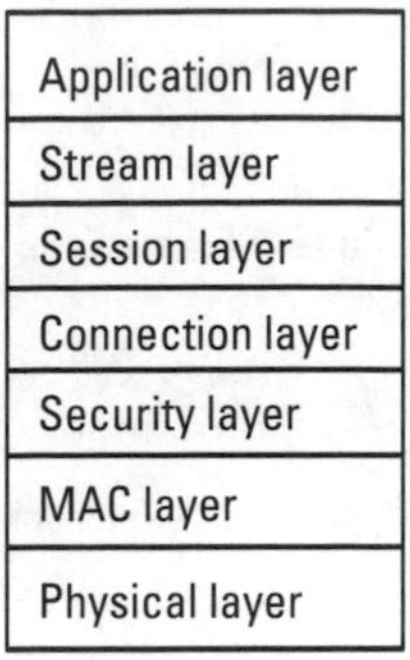

Application layer
Stream layer
Session layer
Connection layer
Security layer
MAC layer
Physical layer

Figure 4.5 1xEV-DO protocol stack. (*After:* [7] © 3GPP2.)

The application layer contains a default signaling protocol for use between devices supporting the same packet protocols, a default packet application, and potentially up to two other applications. Each of these applications generates its own stream, so that there can be up to four streams, each with its own QoS and headers, of which stream 0 supports the default signaling and stream 1 supports the default packet application. The session layer comprises session management and configuration as well as address management, while the connection layer handles idle mode, initialization, routing updates, and connection. Security covers protocol, key exchange, encryption, and authentication. The MAC layer supports forward and reverse traffic channels, a control channel, and an access channel. The radio link protocol supports retransmission and duplicate detection for streams of octets. If the packet protocol is IP, then the RLP is PPP as in basic cdma2000. Mechanisms for QoS control are not defined in general, but the MAC control channel includes a data rate control channel that enables a terminal to control the transmission rate on a traffic channel when in the variable rate state. There are 13 valid DRC values for the forward traffic channel covering rates from 0 to 2,457.6 Kbps, subject to specific slot lengths for the packet. There is also a fixed rate state for data transmission with constant rate, and standard messages for transition between these two states. On the reverse traffic channel, rates are restricted to 0, 9.6, 19.2, 38.4, 76.8, or 153.6 Kbps.

1xEV-DV is intended to support both voice and data and to be compatible with cdma2000 1X and 3X, and optionally able to interwork with 1xEV-DO. Its aims include the support of multiple traffic types with different QoS constraints on a single radio channel. The intended rates include simultaneous peaks of 2.4 Mbps forward and 2 Mbps reverse for pure packet data in pedestrian or stationary environments, and corresponding values of 2.4 Mbps and 1.25 Mbps for high-speed vehicles and an average of 600 Kbps in either direction.

References

[1]   Third Generation Partnership Project 2, S.R0003.A, "System Capability Guide— Release B," http://www.3gpp2.org.

[2]   Third Generation Partnership Project 2, S.C0003.A, "Medium Access Control (MAC) for cdma2000 Spread Spectrum Systems—Release A," http://www.3gpp2.org.

[3]   Third Generation Partnership Project 2, S.C0004.A, "Upper Layer (L3) Signaling Standard for cdma2000 Spread Spectrum Systems—Release A," http://www.3gpp2.org.

[4]   Third Generation Partnership Project 2, S.C0005.A, "Upper Layer (L3) Signaling Standard for cdma2000 Spread Spectrum Systems—Release A," http://www.3gpp2.org.

[5]   Third Generation Partnership Project 2, S.A0002.A, "Physical Layer Standard for cdma2000 Spread Spectrum Systems," http://www.3gpp2.org.

[6]   Third Generation Partnership Project 2, P.S0001-A, "Wireless IP Network Standard," http://www.3gpp2.org.

[7]   Third Generation Partnership Project 2, S.C0024, "cdma2000 High Rate Packet Data Air Interface Specification," http://www.3gpp2.org.

[8]   Third Generation Partnership Project 2, S.R0026, "High-Speed Data Enhancements for cdma2000 1X—Voice and Data Integration," http://www.3gpp2.org.

5

GPRS

5.1 General Principles

The step from GSM straight to full WCDMA requires a full network upgrade because of the change from TDMA to CDMA access, and this would be very expensive and technologically risky before an extensive market exists. There are two basic ways of increasing data rates while retaining GSM's TDMA architecture. The first is employed by GPRS and uses multiple timeslots. The second is to use an increased modulation rate, and EDGE combines this with GPRS as EGPRS. As a result, many GSM operators implement 2.5G solutions based on GPRS or EGPRS as an initial step to provide reasonable performance for packet data with gradual evolution to WCDMA using multifunctional handsets. GPRS uses the original GSM equipment with the addition of specific gateway nodes, so it is fairly easy for an operator to provide the same geographic coverage for GPRS as for GSM. In the United States there is an analogous path from the old TDMA AMPS service to IS-136 TDMA service, but the main evolutionary path from AMPS has been the transition to IS-95 CDMA. The IS-95 CDMA standards have a much smoother path to cdma2000 without the need for GPRS. IS-95B (cdmaOne) already has an IP stack built into the handsets and should be capable of operating at up to 114 Kbps on both IS-95B itself and subsequent cdma2000 networks, thus giving comparable functionality to GPRS. The base stations for CDMA should only require new software and channel cards for upgrade to cdma2000 1X and 3X, while 1X and 3X handsets should be capable of operating on earlier versions of the network at the old level of functionality. For WCDMA and GPRS it seems likely that handsets will have dual functionality to meet the separate requirements of the two network standards.

The main function of GPRS is to provide medium-speed data services, but the specifications also allow for the possibility of supporting voice group and broadcast services. GPRS has been defined in two stages: stage 1 was only for GSM [1], while stage 2 has both GSM and UMTS interfaces [2]. This chapter refers to stage 2, but mainly with the GSM features. Three operational modes are defined for GPRS mobiles:

- *Class A:* The mobile is able to support simultaneous transfer of both PS and CS traffic using different timeslots of GSM TDMA.

- *Class B:* The mobile is able to attach to both PS and CS services at the same time and is able to receive pages for both types, but only to transfer traffic of one type at a time. It should be able to suspend data transfer while accepting a CS call, then resume transfer after termination of the latter.

- *Class C:* The mobile can only handle packet traffic.

A further distinction between mobile types is the ability to support COMPACT operation. This makes use of multiple radio frequencies within a cell instead of using only one. Most early GPRS handsets belong to class B, they offer a maximum of three or four timeslots and do not support COMPACT.

The ability of GPRS mobiles to make use of multiple slots also varies, and this leads to the definition of a multislot class. This is based on the numbers of slots that can be used for transmission and reception, with a further distinction based on the time required to get ready to receive or transmit. Multislot class numbers range from 1 (for a mobile able to use only a single slot) up to 29 (for a mobile able to receive or transmit on all eight slots). GPRS mobiles that cannot receive and transmit at the same time are said to be of type 1, while those that can are type 2.

A GPRS mobile should eventually be able to operate in either GSM mode or UMTS mode, but initially only GSM mode is available, and that is the mode described in this chapter. The network architecture is shown in Figure 5.1.

Voice traffic goes to the MSC as in GSM. The mobile station (MS) is indicated as the union of a user's terminal equipment (TE) and the mobile terminal (MT) itself, although it may be a single device. The initial interface

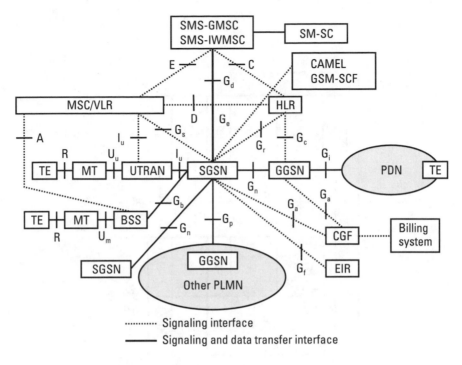

Figure 5.1 GPRS network architecture. (*Source:* [2] © ETSI 2001.)

via GSM uses the base station system (BSS), while the later interface into the 3G network uses the UTRAN connections. Home location register (HLR), visitor location register (VLR), and equipment identification register (EIR) indicate home, visitor, and equipment identification as in GSM, while SMS denotes the GSM short messaging service. The key GPRS additions are the support GPRS service node (SGSN) and the gateway GPRS support node (GGSN) that are usually IP routers with additional specialized software. There is a significant functional difference between the GSNs for 2.5G and 3G systems, so a prefix 2 (as in 2-SGSN) indicates that they are for interface to the BSS of a 2G network, while 3-SGSN would indicate full UMTS operation with a different physical layer. The BSS is itself essentially a 2G BTS that has been upgraded with a packet control unit (PCU) and new software.

In terms of the interfaces shown in Figure 5.1, this chapter is primarily concerned with U_m and G_b from MS to BSS and BSS to SGSN, respectively. The overall structure of the GPRS protocol stack for GSM mode is shown in Figure 5.2.

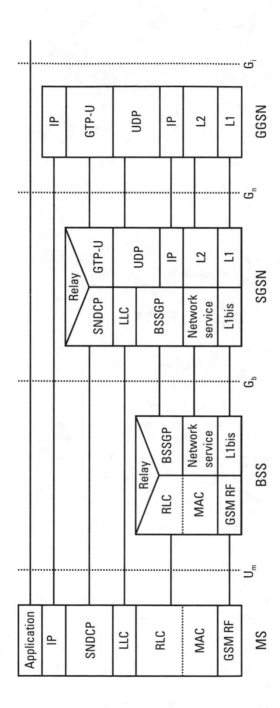

Figure 5.2 GPRS protocol stack. (*Source:* [2] © ETSI 2001.)

In the case of UMTS, the operation of the mobile uses UMTS functions at the U_u interface (see Chapter 3). This chapter covers the protocols that are used between the MS, BSS, and SGSN. The GPRS Tunneling Protocol (GTP) [3] at interface G_n is described in Section 8.1.3.

Mobility Management and Addressing

A mobile can exist in one of three MM states for GSM:

- *Idle:* In this state the mobile is not attached to GPRS mobility management and is unreachable, but it is able to choose a PLMN and select or deselect a cell.

- *Standby:* This state is reached from the idle state by the mobile performing a GPRS attach. In this state it is able to receive paging information and requests for both GPRS and CS operation. It is also able to initiate PDP context activation or deactivation. It moves from standby state to ready state upon data transmission.

- *Ready:* In this state the mobile is able to send or receive data as well as activating or deactivating PDP contexts. If a ready state activity timer expires, it moves back to standby and then to idle if it sends a GPRS detach.

Information about the mobile is stored in the HLR with which it is registered and also by the SGSN. The MS has to identify itself in the GPRS attach and does so by means of either its IMSI or a TMSI. The IMSI consists of a three-digit mobile country code (MCC), a two- or three-digit mobile network code (MNC), and a mobile subscriber identity number (MSIN) within the network, so that the total IMSI does not exceed 15 digits [4]. A TMSI is a 32-bit local identifier allocated by either the VLR or SGSN to which the MS is temporarily attached. A TMSI is used for both circuit and PS services, and these are distinguished by the first 2 bits, where 11 indicates the PS case. A VLR may also allocate a four-octet local mobile subscriber identity (LMSI). An MS that is attached to the PS services is also assigned a three-octet packet TMSI (P-TMSI) by the SGSN. In addition, the MS or SGSN assigns a temporary logical link identifier (TLLI) on the basis of the P-TMSI (if there is one) or according to a specific set of rules [4] for use in the logical link control (LLC) (see Section 5.2.3) and RLC/MAC layer.

Its home PLMN (HPLMN) operator also assigns the MS an IP address. This can either be a permanent static address for IPv4 or IPv6 or a dynamic address. Similarly, a visited PLMN (VPLMN) can provide IP addresses. The choice of whether static or dynamic addresses are to be used is made by the HPLMN operator.

Each GPRS support node (GSN) also has an address. This consists of a 2-bit type field with values 0 and 1 for IPv4 and IPv6, respectively, an IP address length field, and the actual IP address. An access point name (APN) is also assigned to a GGSN and consists of a mandatory APN network identifier to identify the external network to which the gateway is attached and an optional APN operator identifier to show which PLMN it belongs to. The APN is resolved to the IP address by the GPRS domain name server (DNS).

GPRS uses two types of PDP: IP itself and Point-to-Point Protocol (PPP) [5]. PDP contexts are established between the MS and network for specific PDP types, addresses, QoS requirements, and network service access point identifier (NSAPI) (see Section 5.2.3). A single PDP address may have more than one context (e.g., for different QoS requirements) in Release 4 (but not in the 1997–8 version of GPRS), and if so, uses a traffic flow template (TFT) [2, 6] to identify all but at most one of the contexts. The TFT consists of one to eight different packet filters for the PDP address. They are defined by one or more of the following parameters: source address and subnet mask, IPv4 protocol number, source or destination port number, IPSec security parameter, IPv4 type of service (TOS) (see Section 7.4.1), or IPv6 (see Section 7.4.2) alternatives to the IPv4 parameters. Of these options, IPv4 TOS gives a measure of QoS requested but is liable to be used in inconsistent ways (see Sections 7.4.1 and 7.4.4).

Once the MS is in the ready state, PDP contexts may be activated, deactivated, or modified by either the MS or the network. PDP activation involves the BSS, SGSN, and GGSN in the sequence shown in Figure 5.3 for the case of activation by the MS.

A secondary PDP context activation with different TFT can be used for different QoS traffic for the MS without deactivating the first context, but this option is not available on early handsets. The GGSN compares incoming traffic with the active TFTs and assigns it to the channel for the correct TFT; it drops nonconforming traffic unless there is an active context without TFT.

5.2 Layer 3

5.2.1 Radio Resource Management

The general purpose of radio resource (RR) procedures [7–9] is to establish, maintain, and release RR connections that allow a point-to-point dialog

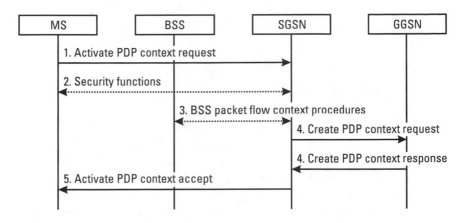

Figure 5.3 PDP activation. (*Source:* [2] © ETSI 2001.)

between the network and a mobile station. This includes the cell selection/reselection and the handover procedures. Moreover, radio resource management procedures include the reception of the unidirectional BCCHs and CCCHs when no radio resource connection is established. This permits automatic cell selection/reselection.

If the voice group call services (VGCS) listening or voice broadcasting services (VBS) listening are supported, the radio resource management also includes the functions for the reception of the voice group call channel or the voice broadcast channel, respectively, and the automatic cell reselection of the mobile station in group receive mode.

If VGCS talking is supported, the radio resource management also includes the functions for the seizure and release of the voice group call channel.

If GPRS point-to-point services are supported, as will be the case here, the radio resource management procedures include functions related to the management of transmission resources on packet data physical channels. This includes the broadcast of system information to support a mobile station in packet idle and packet transfer modes.

A mobile can be in one of seven potential modes:

1. Idle—the mobile has no RR connection;

2. Dedicated mode—the mobile has a point-to-point bidirectional RR connection and a data link connection on the main DCCH;

3. Group receive mode;

4. Group transmit mode;

5. Packet idle mode;

6. Packet transfer mode;

7. Dual transfer mode.

States 1 and 2 describe the state of CS connectivity of an MS. Modes 3 and 4 are not discussed in this chapter since the main concern here is with data applications. Modes 5 and 6 describe GPRS operation. In the former there is no packet transfer, while in the latter the MS has been assigned a temporary block flow to send packets. Mode 7 is only applicable to mobiles that belong to class A, and it describes the state in which such a mobile has both a CS connection (dedicated mode) and makes a successful packet request to send data.

5.2.2   Subnetwork Dependent Convergence Protocol

The Subnetwork Dependent Convergence Protocol (SNDCP) [10] and LLC [11] are situated between the IP network level and RLC layers logically and within the MS and SGSN physically, with SNDCP being above LLC. It interacts with the network traffic above, and the session management (SM) and the LLC below, as indicated in Figure 5.4. It carries user traffic only, while MM traffic goes direct to the LLC.

The SNDCP has four functions:

1. Multiplexing of data for different PDPs;

2. Compression/decompression of data payloads;

3. Compression/decompression of headers;

4. Segmentation/reassembly of network protocol data units (N-PDU) for the underlying logical link.

Interaction with the LLC takes place via service access points, with the NSAPI defining an index into the PDP context of the SNDCP user and the service access point identifier (SAPI) defining the entry point into the LLC. There are different NSAPIs for IPv4 and IPv6. Session management tells the SNDCP via an SNSM-ACTIVATE indication when a NSAPI is to be activated for data transfer and includes information on the QoS negotiated and the SAPI to use. The choice of NSAPI is originally made by the mobile at the time of PDP context activation and passed by SM to its own SNDCP. Each

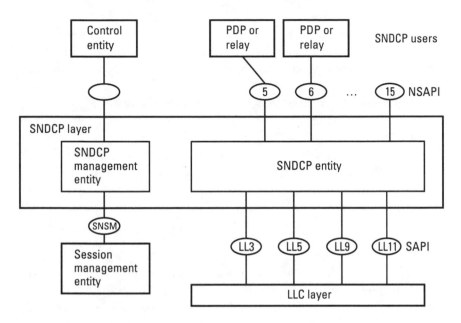

Figure 5.4 SNDCP. (*Source:* [10] © ETSI 2000.)

SAPI has a specific QoS profile, and may be shared by multiple NSAPIs that are compatible with the profile.

The main primitives between the SNDCP and the LLC are LLC-Establish to set up or reestablish a logical link for a specific SAPI and two data transfer commands, the LLC-DATA Request to send an N-PDU in acknowledged mode, and the LLC-UNITDATA Request to send the N-PDU in unacknowledged mode. These two requests each contain QoS information. The choice of QoS information depends on the direction (i.e., whether from the mobile or from the SGSN). The mobile includes the peak throughput required and a radio priority for the RLC/MAC layer, while the SGSN gives features such as precedence class, delay class, and peak throughput, such as might be provided from an IP network connection.

The SNDCP may multiplex traffic from multiple NSAPIs for transmission over the logical link for a suitable SAPI.

The SNDPC can negotiate payload compression using ITU V.42bis or V.44, although this feature is optional and so is not included in all implementations. If present, it is put in a DCOMP field in the LLC-ESTABLISH or LLC-XID, with 0 indicating V.42 bis and 1 for V.44.

The ability to negotiate header compression is mandatory, and there are two options at present that go in a similar PCOMP field. Value 0 corresponds to the use of RFC1144 for TCP/IP header compression, and value 1 corresponds with RFC2507 (see Section 7.4.10 for a description of these). Each of these options contains several additional parameters that have to be negotiated.

Segmentation has to be performed to ensure that the N-PDU does not exceed a maximum length N201 (octets); splitting the N-PDU into multiple SN-PDU segments with an M-bit more flag set is necessary if this happens. The value of N201 may differ for acknowledged and unacknowledged modes with values N201-I and N201-U, respectively. A flag *F* also has to be set if PCOMP or DCOMP are used in either case and also if a sequence number is included for the acknowledged case.

5.2.3   LLC Operation

The LLC [11] runs between the mobile and SGSN to provide a reliable link and is based on both Link Access Protocol D (LAPD) and the CS RLP [12]. It supports data confidentiality via ciphering, different levels of data priority, and the use of the same radio resources by multiple mobiles. It operates above the RLC layer in the MS and above the BSSGP in the SGSN.

An LLC layer connection is defined by its SAPI and the TLLI for the MS; together these form the data link connection identifier (DLCI). The TLLI is provided by GPRS mobility management (GMM) in a LLGMM-ASSIGN-Request. The various SAPI for different services are shown in Figure 5.5, where SMS is short message service and TOM is tunneling of messages (between MS and SGSN).

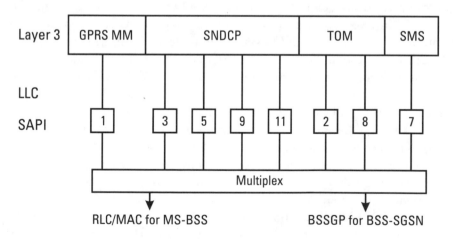

Figure 5.5   SAPI. (*After:* [11] © ETSI 2001.)

The principle functions of the LLC are as follows:

- Provision of connections identified by their DLCIs;
- Sequence control of frames;
- Error detection;
- Recovery from errors where appropriate;
- Notification of unrecoverable errors;
- Flow control where appropriate;
- Ciphering.

The LLC provides both an acknowledged and an unacknowledged service to both the RLC and BSSGP. Error recovery and flow control only apply to the acknowledged service. MM messages are carried by the unacknowledged service. The unacknowledged service has two modes: protected and unprotected, according to whether frames with errors are dropped or retained, respectively.

The SAPI defines the first 4 bits of the data frames, so 16 values are possible. Of these, value 1 defines the SAPI for GMM, while 3, 5, 9, and 11 are the values for user data DLCIs. Frames contain a control field, up to N201 octets of data for I-frames, and a three-octet frame check sequence (FCS) field. The FCS is determined by a CRC calculation for the header and the full N201 data octets, except in the case of unacknowledged UI frames in unprotected mode where only a smaller number (N202) of data bytes are covered. A flag indicates protection, and the advantage of unprotected mode is that some frames, such as those for AMR speech, contain data of differing levels of criticality. If an error occurs in the unimportant later parts of the frame, then it is not worth indicating the error.

The values of N201-I and N201-U are negotiable between the MS and SGSN for each SAPI within 140 to 1,520 octets in general, but with defaults that are equal to 500 octets for unconfirmed information (UI) and 1,503 for I-frames for the data SAPIs 3, 5, 9, and 11.

Flow control is provided for I-frames using sequence numbers, windowing (based on buffer size), acknowledgments, and retransmission timers. GRR-DATA-REQ/IND are used for acknowledged LL-PDUs between the LLC and RLC or vice-versa, and GRR-UNITDATA-REQ/IND are used for unacknowledged LL-PDUs. These include QoS information such as peak throughput and radio priority. Only UIs for SAPIs 3, 5, 9, and 11 use the UNITDATA primitives.

Similarly, the LLC-BSSGP use BSSGP-DL-DATA and BSSGP-DL-UNITDATA primitives.

5.2.4   BSS-SGSN Protocol

The BSS-SGSN Protocol (BSSGP) [13] that is used between the BSS and SGSN sits between the LLC and network service layers in the SGSN and interfaces to the RLC of the BSS via a relay function. The network service is not specified but is typically IP or frame relay. The primary function of the BSSGP is the transfer of radio information for the RLC/MAC between BSS and SGSN in either direction. In particular, it supports the transfer of LLC-PDUs between MS and SGSN and provides flow control for this purpose. Communication between the BSSGP entities (SGSN and BSS) takes place on either a point-to-point (PTP) or point-to-multipoint (PTM) basis using BSSGP virtual connections (BVC) over the network, each of which has its own identifier (BVCI). Data transfer between the MS and SGSN is PTP, while some BSS management functions are PTM. PTM BVCI are configured statically and PTP dynamically at the SGSN, while at the BSS the PTP BVCI can be either static or dynamic.

The SGSN provides the BSS with information about MS data transfers that is stored in BSS contexts. Each BSS context refers to a single MS but may contain several distinct packet flows, each identified by its own packet flow identifier (PFI). The PFI is needed if the optional packet flow context (PFC) is supported. If the PFI is not stated, then packet flow is handled on a best-efforts basis.

Packet flow control is only provided on the DL. The BSS has a buffer for each BVC and indicates to the SGSN via FLOW-CONTROL-PDUs the maximum flow it can take together with the maximum rate for a specific MS and TLLI. This is based on a leaky bucket approach, and the information sent consists of the bucket size (B_{Max}), bucket leak rate (R), and bucket full ratio for the BVC or for an MS. An LLC-PDU is only sent if it is compatible with both of these flow control measures [i.e., if the current bucket count (B) plus the LLC-PDU length does not exceed either B_{Max}]. Rules are provided for calculation of the MS bucket size, and there is also a default. There is also a minimum interval for FLOW-CONTROL-PDUs in normal conditions. The peak rate quoted in the QoS profile (see Sections 5.5 and 9.2) is related to the bucket leak rate (R). Precedence in the QoS profile is the radio priority (1–4) for the UL-UNITDATA and high, medium, or low (coded as 0, 1, 2) for the DL.

5.3 Layer 2 RLC and MAC

Layer 2 consists of the RLC and the MAC [7] beneath it, which are both distinct from those described in Chapter 3 for UMTS. RLC communicates with the RRC layer and upper layer signaling and LLC user traffic, while the MAC interfaces to the physical layer and controls access to the radio interface. Most pure signaling traffic bypasses this sublayer. A separate protocol, BSSGP, is used at this level on the link from the BSS to the SGSN. The structure of the RLC/MAC layer and its position in relation to the LLC, MM, and physical layers is shown in Figure 5.6.

As indicated, the control channels use the data link layer [14] (which uses 21 or 23 octet blocks), while the packet data uses the RLC/MAC. As

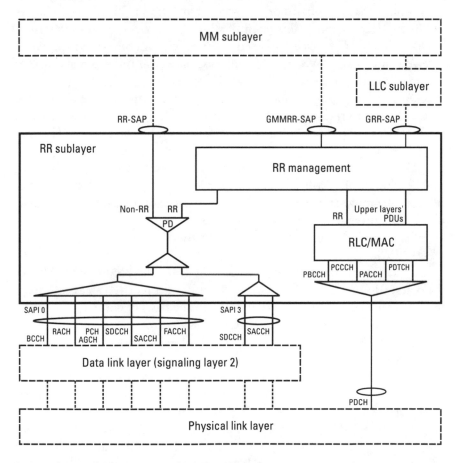

Figure 5.6 RLC/MAC. (*Source:* [7] © ETSI 2001.)

regards QoS, the latter is much more significant and is described in the following sections. The permitted combinations of logical channels over the physical channels are specified in [15].

5.3.1 RLC

The prime function of the RLC is to segment and reassemble logical link control PDUs into RLC/MAC blocks [7]. It has two modes of operation: AM and UM. In AM the RLC is also responsible for sequencing, acknowledgment, and retransmission of blocks to provide backward error correction. In EGPRS AM it also provides incremental redundancy that avoids resegmentation for ARQ retransmissions.

Packet transfer is based on temporary block flows (TBF) that each have two peer RLCs as end points, each of which has a receiver and a transmitter for RLC/MAC blocks. Each end-point RLC has both a transmit and a receive window size (WS) for these blocks. Flow control also makes use of a sequence number space (SNS) and a block sequence number (BSN). In RLC AM each end point maintains receive and transmit state variables $V(Q)$ and $V(S)$, respectively, as well as an acknowledge state variable $V(A)$. In RLC UM the sole use of the BSN is for reassembly of LLC PDUs from RLC/MAC blocks.

$V(S)$ denotes the BSN of the next block to be transmitted, and it is set to 0 at the start of a TBF and incremented by 1 with each block transmitted. It lies in the range 0 to SNS-1 and must not exceed the value of $V(A)$ modulo SNS by more than the window size WS.

The receive state $V(Q)$ must satisfy the inequality $V(Q) \leq \text{BSN} \leq V(Q) + \text{WS}$.

The transmitter maintains the value $V(A)$ for the BSN modulo SNS of the oldest RLC/MAC block for which it has not yet received an acknowledgment from its peer. This value is updated on the basis of data in the received block bitmap contained in ACK/NACK messages from the peer.

The ACK/NACK message is sent in both RLC AM and UM but may be ignored in the latter. It consists of a starting sequence number (SSN) and the received block bitmap (RBB). Both the receiver and transmitter maintain copies of the RBB. The receiver sets the SSN to the value of the current receive state and makes an array of WS entries in which it sets each entry to 1 for a valid received block and to 0 for an invalid block. In addition to updating $V(A)$, the transmitter also uses the RBB to maintain an array of ACKed/NACKed states for each block sent in a window size, setting elements to ACKed for 1 in the RBB and NACKed for 0.

In GPRS the window size WS is 64, and the SNS is 128. For EGPRS these values are extended, with SNS rising to 2,048 and WS ranging from 64 up to a maximum depending on the number of slots used (from 1 to 8) with a peak of 1,024 for eight slots. There are also minimum window sizes for EGPRS. Table 5.1 shows the maxima and minima in relation to the multislot capabilities of the mobile.

In all cases the receiver and transmitter have to maintain a large number of timers to control retransmissions and request/release of temporary block flows.

Logical Channels

Logical channels comprise both CS traffic channels and packet data traffic channels (PDTCH). Traffic channels [16] carry either voice or data in CS mode and consist of three basic types:

1. Full-rate traffic channel (TCH/F) with gross information rate of 22.8 Kbps;
2. Half-rate traffic channel (TCH/H) with gross information rate of 11.4 Kbps;
3. Enhanced full-rate traffic channel (E-TCH/F) with gross information rate of 69.6 Kbps. The distinction between E-TCH/F and the others is that the modulation scheme used in EDGE is 8-PSK instead of GMSK.

Table 5.1
EGPRS Window Sizes

Multislot Capability	Minimum	Maximum
1	64	192
2	96	256
3	160	384
4	160	512
5	224	640
6	320	768
7	352	896
8	512	1,024

When used for data, the actual data rates and channel acronyms are as given in Table 5.2.

The difference between these data rates and the three possible gross information rates is primarily due to the need for channel coding to provide error correction together with some small overheads. Layer 2 uses a wide range of distinct logical channels to carry out its PS needs. These are as follows:

- Common packet control channels (PCCCH and CPCCCH), which include packet random access channels (PRACH and CPRACH), packet paging channels (PPCH and CPCCH), packet access grant channels (PAGCH and CPAGCH), and packet notification channels (PNCH and CPNCH);
- Packet broadcast control channels (PBCCH and CPBCCH);
- Packet data traffic channels (PDTCH);
- Packet associated control channel (PACCH);
- Packet timing advance control channels (PTCCH).

Common packet control channels are a generic term for the four subtypes PRACH, PPCH, PAGCH, and PNCH together with their compact versions for mobiles able to support that mode.

PRACH and CPRACH are used by the MS to initiate UL data transfer or signaling, while the paging channels are used by the network to do this on the DL. PAGCH and CPAGCH are used in the packet transfer

Table 5.2
Data Rates

Channel Name	Data Rate (Kbps)
TCH/F14.4	14.4
TCH/F9.6	9.6
TCH/F4.8	4.8
TCH/F2.4	≤ 2.4
TCH/H4.8	4.8
TCH/H2.4	≤ 2.4
E-TCH/F28.8	28.8
E-TCH/F32.0	32.0
E-TCH/F43.2	43.2

establishment phase to grant resources to the MS. PNCH and CPNCH notify the MS of a point-to-multipoint packet transfer, while the broadcast channels are used to transmit system information.

The PDTCH are allocated to either a single MS or to a group of MSs for the purpose of packet data transfer. In multislot operation a single MS may be allocated several PDTCH for use in parallel to achieve a higher data throughput. PDTCH are unidirectional for either UL or DL and can be either full-rate or half-rate depending on the physical data channel to which they are mapped in layer 1.

The remaining two control types are dedicated to a specific MS. PACCH carries signaling, power control information, and resource assignments to a specific MS, while PTCCH is for timing control information.

5.3.2 MAC-MODE

The MAC layer controls the allocation of channels and timeslots. The GPRS MAC function is responsible for providing:

1. Efficient multiplexing of data and control signaling on both UL and DL, the control of which resides on the network side. On the DL, multiplexing is controlled by a scheduling mechanism. On the UL, multiplexing is controlled by medium allocation to individual users (e.g., in response to service request);
2. Mobile originated channel access and contention resolution between channel access attempts, including collision detection and recovery;
3. Mobile terminated channel access and scheduling of access attempts, including queuing of packet accesses;
4. Priority handling.

This is performed through the exchange of control messages between the network and mobile. The messages most directly associated with user traffic are channel, packet channel, and resource reallocation messages from the mobile, and packet UL/DL assignment messages from the network.

There are five modes of operation of the MAC layer, and these are selected by the network and passed to the mobile in the packet assignment messages:

1. Dynamic;
2. Extended dynamic;

3. Fixed full-duplex;

4. Fixed half-duplex;

5. Exclusive.

Dynamic assigns a single packet dedicated channel (PDCH) to a mobile that is already in either packet transfer or dual transfer mode. Extended dynamic assigns up to eight PDCHs to a mobile in either of these modes according to the mobile's multislot capability. These two allocation modes provide for sharing of resources based on the USF field at the start of each radio block. Fixed reserves part of the UL PDCH for a specific mobile for a certain period. Exclusive is only applicable to mobiles that are in the dual transfer state.

The assignments are not permanent, and they provide a TBF and associated flow identity (TFI). The TFI is assigned in a resource assignment message immediately prior to the transfer of LLC frames for that TBF to or from the MS. It is a binary number in the range 0 to 31 that is used instead of the MS identity by the RLC/MAC. The TBFs can be either close-ended or open-ended, where the former specify transmission of not more than a fixed number of RLC blocks (plus any necessary control blocks and retransmissions) and the latter do not impose limits. The packet assignments also specify the RLC mode that is to be used, a starting frame number, and in the case of EGPRS the window size. The typical order of magnitude for the duration of an assignment is about a second. Open-ended is preferable for real-time traffic to prevent interruptions in flow.

5.3.3 BSS-SGSN Network Service

Network service is the bottom part of layer 2 of the BSS to SGSN connectivity [17]. It uses either frame relay or IP subnetworks over a variety of possible physical links. The basic features are network service virtual links (NS-VL) and network service virtual connections (NS-VC). An NS-VL is a virtual communication path between the BSS or SGSN and intermediate network, or between BSS and SGSN if connected point to point. An NS-VC is an end-to-end virtual connection between pairs of NS entities. Each physical link supports one or more NS-VL. For frame relay there is one-to-one correspondence between NS-VC and NS-VL, but for IP one NS-VL may have several NS-VCs. NS-VCs between the same pair of NS entities constitute a single NS-VC group. User traffic for a given BVC is sent in unacknowledged frames at this level with load sharing between unblocked NS-VCs in the group.

5.3.4 Radio Link Protocol

This is the protocol that is used to support data traffic, such as fax or modem links, over the CS channels [12]. There are three versions: 0, 1, and 2, supporting single link, single link plus compression, and multilink (up to eight links), respectively. It uses a version of HDLC protocol and sets asynchronous balanced mode (ABM) when the data connection is active or asynchronous disconnected mode (ADM) otherwise. It uses frames of 576-bit length for UMTS operation and 240 for GSM. Handovers where the channelization code changes require resynchronization with interruption to data flow, and handovers between GSM and UMTS need remapping.

5.4 Layer 1

GPRS uses the same physical layer as GSM; hence, it is based on 200-KHz carriers, each of which is subdivided into eight channels that are arranged as timeslots, but whereas in GSM a user is assigned a single timeslot in each direction, a GPRS user can have multiple (usually unequal) numbers of timeslots in each direction up to the maximum of eight depending on multislot class. HSCSD also uses up to eight channels of this type, but it assigns them for the duration of a call and they are subject to the additional restriction that all slots must use the same channel mode.

Physical channels are defined as sequences of radio frequency channels and timeslots. Each TDMA frame consists of eight timeslots, and a physical channel always uses the same timeslots in each TDMA frame.

The number of physical channels available depends on the frequency band being used. The norm is either GSM 900 or GSM 1,800, but others are occasionally used, and the numbers of channels are as follows: 35 (GSM 450), 35 (GSM 480), 74 (GSM 700), 124 (GSM 850), 194 (GSM 900), 374 (DCS 1,800), and 299 (PCS 1,900) radio frequency channels, with a guard band of 200 kHz at each end of the subbands.

EGPRS offers 8-PSK modulation as an alternative to the usual GMSK to allow higher data rates. It also provides incremental redundancy to allow higher rates of ARQ retransmission by avoiding the resegmenting of RLC data blocks that would otherwise be the norm.

GPRS and EGPRS mobiles can share the same physical channels.

Layer 1 functionality is split between two sublayers, the physical link layer and the RF-layer. The physical link layer interfaces to the RLC/MAC layer above and to the RF-layer beneath. It is responsible for providing a physical channel between the network and the mobile. Its functions include channel coding to provide forward error correction, the interleaving of a single radio block over four TDMA frames, and the detection of network congestion. It also performs specific control functions including synchronization, link quality monitoring, cell selection/reselection procedures, and power control.

The RF-layer provides modulation and demodulation of bit streams from the physical link layer and uses GMSK for GPRS and EGPRS, with the option of 8-PSK for EGPRS. There are detailed tables to define how CS traffic is carried over physical channels. There is only one type of physical channel for GPRS/EGPRS packet data, and that is the PDCH. A single PDCH is a shared medium that supports multiple logical channels and multiple mobiles (except in dual transfer mode) and only carries PS data. A PDCH is either full-rate (PDCH/F) or half-rate (PDCH/H), but the latter is only applicable to dual transfer mode.

The UL and DL channels are asymmetric and operate independently. Contention resolution between different mobiles is required on the UL, but not on the DL where transmission is directly controlled by the network. Multiplexing of radio blocks from different mobiles is controlled through the use of the UL state flag (USF) on the DL. The USF is a 3-bit field in the MAC header at the start of each radio block on the DL that supports eight different values to control the multiplexing. Its usage differs according to whether the PDCH carries a PCCCH or not. Where there is no PCCCH, the eight USF values are used to reserve the UL for different mobiles, but if the PCCCH is carried, then one of the USF values is used for the PRACH instead of a mobile. When GPRS and EGPRS mobiles share the same PDCH, the EGPRS mobiles are restricted to the use of the standard GMSK modulation only.

Radio Blocks

For GPRS, a radio block for data transfer consists of one MAC header, one RLC header, and one RLC data block [8]. It is always carried by four normal bursts, as shown in Figure 5.7.

For EGPRS, a radio block for data transfer consists of one RLC/MAC header and one or two RLC data blocks. It is always carried by four normal bursts. The interleaving depends on the channel-coding scheme used. An example of this is shown in Figure 5.8 for the case of EGPRS MCS-8 coding.

The physical link layer is responsible for putting the BCS on the radio block, although it is the RLC that checks it for backward error correction.

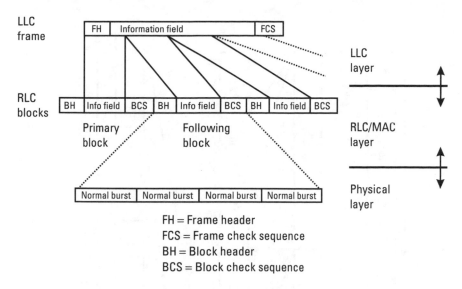

FH = Frame header
FCS = Frame check sequence
BH = Block header
BCS = Block check sequence

Figure 5.7 RLC block forwarding. (*Source:* [8] © ETSI 2001.)

Channel Coding

The coding schemes vary according to whether the MS supports GPRS or EGPRS. For GPRS there are four possible coding schemes, CS-1 to CS-4, that may be used on the packet data traffic channels; the PRACH must always use CS-1.

In the case of EGPRS there are nine different modulation and coding schemes, MCS-1 to MCS-9, all of which are mandatory on the DL, but only MCS-1 to MCS-4 are mandatory on the UL unless the MS supports 8-PSK modulation.

CS-1 to CS-3 use rate 1/2 convolution coding (see Section 2.5), while CS-4 is unencoded. CS-2 and CS-3 are punctured versions of CS-1 that give effective code rates of about 2/3 and 3/4, respectively, thereby permitting more raw data in a radio block and thus providing effective transmission rates of about 13.4 Kbps and 15.6 Kbps, as compared to 9.05 Kbps for CS-1 and 21.4 Kbps for CS-4.

MCS-1 to MCS-9 can be grouped into three families depending on the size of their payload units. Family A has payload units of 37 or 34 bytes, while families B and C have units of 28 and 22 bytes, respectively. Families A and B can carry 1, 2, or 4 payloads in a radio block, while family C takes only 1 or 2. Family A has MCS-3, -6, -8, and -9; B has MCS-2, -5, and -7; and C has MCS-1 and -4 depending on the number of payload units. MCS-7, -8,

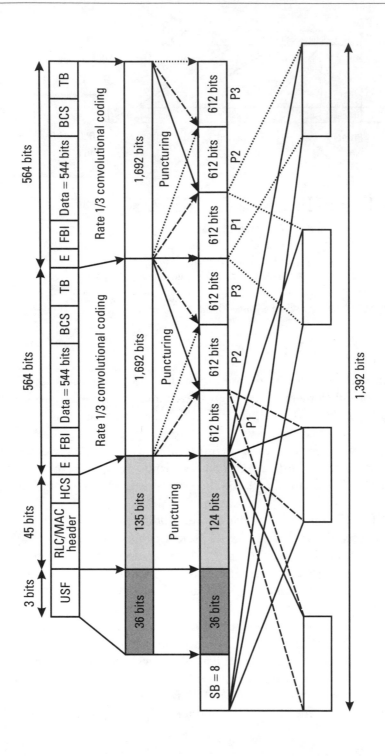

Figure 5.8 MCS-8. (*Source:* [8] © ETSI 2001.)

and -9 support four payload units and convey two RLC blocks in 20 ms, as opposed to 1 RLC block for the remainder. These convolution codes are of rate 1/3. MCS-1 to MCS-4 are used with GMSK modulation, and the remainder are used with 8-PSK. The MCS may be changed within a family from TBF to TBF according to the link quality.

PDCH Multiframe Structure

The PDCH is structured as a 52-frame TDMA multiframe [8] in order to send individual logical channels in a systematic manner. The 52 frames are divided into 12 radio blocks, each consisting of four frames, together with two idle frames and two frames for the timing channel PTCCH as shown in Figure 5.9.

This structuring applies particularly to the DL. One PDCH (which may have PCCCH) is indicated on BCCH, and its first block, plus up to the next three if necessary, are used for PBCCH. Where PDCH contains PCCCH (with or without PBCCH), the next of up to 12 blocks are used for PAGCH, PNCH, PDTCH, and PACCH on the DL, and the remainder can be used for PPCH as well. On a PDCH that does not contain PCCCH, all blocks can be used for PDTCH and PACCH.

5.5 GPRS QoS

GPRS nominally supports the same set of QoS classes as UMTS (i.e., conversational, streaming, interactive, and background) [18]. The first step in assigning quality occurs when the mobile performs its GPRS attach. At this stage it registers its capabilities. Radio and networking capabilities are provided in the MS classmark [19] that is stored by the network in the MM context for the IMSI so long as the MS is attached. The Radio Access Capability element is split into two parts, one for GSM and the other for UMTS. The GSM element contains multislot capability, power class, and technology type

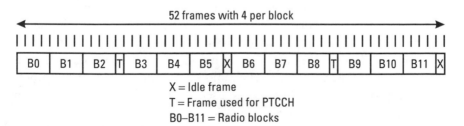

X = Idle frame
T = Frame used for PTCCH
B0–B11 = Radio blocks

Figure 5.9 TDMA multiframe. (*Source:* [8] © ETSI 2001.)

(e.g., GSM 900 or GSM 1,800) for use by the BSS, and it is sent by the MS to the SGSN.

Support of QoS depends on the definition of QoS profiles (see Section 9.2 and [20]) for the services required and on their use on MS-BSS and BSS-SGSN links. The requirements on bit rates, error rates, delays, and ordering are included in a 13-octet QoS information element [6] and carried in the QoS profile [3, 20], which is potentially up to 255 octets in length. The QoS profile registered, requested, and negotiated are separately maintained in an active PDP context with the TFT, which caters for up to nine different bit rates (from 8 to 2,048 Kbps in powers of two). The first GPRS handsets are unable to negotiate QoS and use a registered subscription profile (see Section 9.2.2 and [1]). This provides three reliability classes, four delay classes, the GSM radio precedence, and sometimes the throughput, but this feature may be negotiable even for early handsets. The earliest handsets are only able to use the best-efforts delay class, which provides no delay guarantees. The throughput is based on maximum and mean bit rates that are negotiable up to the information transfer rate determined by the channel type and number of timeslots allocated.

The QoS negotiated is used to select one of the four possible SAPIs for LLC user traffic. The features that the GPRS radio link provides are the choice between acknowledged or unacknowledged RLC modes and control of the data rate and radio priority through the TBFs. There are four radio priority levels, numbered 1 to 4. The first is the highest and is automatically applied to all control messages, while 4 is the lowest and serves as the default level. The MAC layer uses this to decide access to the radio interface during periods of congestion. The maximum number of retransmissions for a message depends on its priority level, and the network sets values from the range 1, 2, 4, or 7 to each level.

QoS in the complete GPRS network is controlled in principle by an IP bearer service manager that resides in a gateway node, and potentially also exists in the mobile station, but not for early models of MS. This manager can use the various modes of QoS control that exist in IP. These are integrated services, differentiated services end-node functions, and IP policy enforcement. Of these, the second two options are mandatory for the gateway node, while all are optional for the mobile. The manager interacts with layer 3 to select the most appropriate radio link options. Much of this procedure is dependent on gateways and is discussed further in Section 8.2, with the policy protocols covered in Section 7.4.6.

The optional integrated services (see Sections 7.4.3 and 7.4.4) provide IP signaling between the two managers using either RSVP or LDP to try to reserve bandwidth that matches the rate and error requirements of the call on an individual basis across the network. The mandatory DiffServ (see Section 7.4.5) merely maps a call into the most appropriate of 15 collective queues with different priorities. These features are described in much more detail in Chapter 7.

References

[1]   ETSI TS 122 060, "GPRS Service Description Stage 1," V.4.2.0 (2001), http://www.etsi.org.

[2]   ETSI TS 123 060, "GPRS Service Description Stage 2," V.4.2.0 (2001), http://www.etsi.org.

[3]   ETSI TS 129 060, "GPRS Tunneling Protocol (GTP)," V.4.2.0 (2001), http://www.etsi.org.

[4]   ETSI TS 123 003, "GPRS Addressing," V.4.2.0 (2001), http://www.etsi.org.

[5]   IETF RFC1661, " Point to Point Protocol (PPP)," 1994.

[6]   ETSI TS 124 008, "GPRS Core Network Procedures," V.4.4.0 (2001), http://www.etsi.org.

[7]   ETSI TS 144 060, "GPRS RLC/MAC," V.4.2.0 (2001), http://www.etsi.org.

[8]   ETSI TS 143 064, "GPRS Radio Interface Stage 2," V.4.1.0 (2001), http://www.etsi.org.

[9]   ETSI TS 144 018, "GERAN Radio Resource Control," V.4.6.0 (2001), http://www.etsi.org.

[10] ETSI TS 144 065, "GPRS SNDCP," V.4.1.0 (2001), http://www.etsi.org.

[11] ETSI TS 144 064, "GPRS LLC," V.4.1.0 (2001), http://www.etsi.org.

[12] ETSI TS 124 022, "Radio Link Protocol," V.4.0.0 (2001), http://www.etsi.org.

[13] ETSI TS 148 018, "BSSGP," V.4.2.0 (2001), http://www.etsi.org.

[14] ETSI TS 144 005, "GSM Data Link Layer," V.4.0.0 (2001), http://www.etsi.org.

[15] ETSI TS 144 003, "GERAN Channel Structures," V.4.0.0 (2001), http://www.etsi.org.

[16] ETSI TS 144 004, "GERAN Layer 1," V.4.0.0 (2001), http://www.etsi.org.

[17] ETSI TS 148 016, "BSS-SGSN Network Service," V.4.2.0 (2001), http://www.etsi.org.

[18] Third Generation Partnership Project, TS 23.207, "End-End QoS," http://www.3gpp.org.

[19] ETSI TS 123 107, "QoS Concept and Architecture," V.4.3.0 (2001), http://www.etsi.org.

[20] ETSI TS 124 088, "UMTS Mobile Radio Interface Layer 3 Specification; Core Network Protocols, Stage 3," V.4.4.0 (2001), http://www.etsi.org.

6

RF Design Overview

6.1 Cell-Capacity Basics

Radio access networks consist of cells served by base stations that support the mobiles in their vicinity. Cells may also be subdivided into angular sectors by the use of directional antennae. GSM, along with its derivatives GPRS and EDGE, makes use of frequency division multiple access (FDMA) to divide the bandwidth allocated to an operator into separate carriers, then divides each carrier into eight channels that are arranged by TDMA into timeslots such that a mobile is assigned one (or more in GPRS and EDGE) timeslot in each direction. FDMA suffers from cochannel interference whereby cells suffer interference from other cells that use the same frequency. This necessitates the use of cell clusters with constituent cells operating on different frequencies; each cluster must be large enough not to suffer cochannel interference from other clusters. The number of cells in the cluster is called the reuse factor and is determined by the need for an SIR high enough to achieve an acceptable bit error rate. This is illustrated in Figure 6.1 for a typical hexagonal array of cells. The three different types of shading indicate different carrier frequencies.

Suppose that an operator is assigned bandwidths W for each of the downlink and uplink, and that a carrier occupies bandwidth B_C so that the number of carriers available is W/B_C. If each carrier can support N traffic channels and the reuse factor is U, then the maximum number of channels that can be supported in a cell (ignoring all complications) is

Figure 6.1 Frequency reuse.

$$\text{Max channels} = (W/B_C) \times (N/U) \qquad (6.1)$$

The Erlang theory of the probability of call blocking for a number of call attempts on a given number of channels shows that it is best to arrange the network to have as high a number of channels in a single cell as possible (i.e., for U to be as small as possible). For GSM and its derivatives, U must be at least 3 (as in Figure 6.1) and may be several times as much in unfavorable circumstances, whereas the use of scrambling codes in cdma2000 and WCDMA greatly reduces the effects of cochannel interference and approaches the optimum scenario $U = 1$ in theory. In practice, effects of mobility mean that the scrambling codes are not quite orthogonal, and at high loads several users may temporarily use the same primary scrambling code (see Section 2.9.4) and hence interfere with each other—although multiuser detection can limit the latter. In GSM there are eight channels per carrier and each carrier occupies 200 KHz, so if an operator has a 10-MHz bandwidth available in each direction, the maximum number of full-duplex channels per cell will be equal to or less than 133 in the best case for the technology of $U = 3$. In GPRS a heavy data user can be assigned up to the full eight timeslots (for about a second at a time), so that the corresponding number of simultaneous heavy users per cell would be up to about 16. The GPRS slots are allocated separately for uplink and downlink with much more traffic on the latter normally. In CDMA the calculation of capacity is more complicated because a wide variety of channel types and data rates are employed, and this is outlined in Section 6.5.

In some situations, particularly in urban environments, it is advantageous to use directional antennae in the base station with little overlap to create separate sectors within a cell that are able to reuse the same carrier frequency in CDMA (but not GSM/GPRS or TDMA technologies where separate frequencies are needed) and so increase capacity. If the angular width of a sector is θ and the angular overlap between sectors is Φ, then the capacity gain is given by $2\pi/(\Phi + \theta)$. A fairly typical situation would be an array of hexagonal cells, each split into three angular sectors as in Figure 6.2, thereby increasing capacity by a factor of almost three.

In this diagram the solid arrows indicate the limits of 120° sectors, while the dashed lines show an extension of about 10° overlap to enable handovers between sectors.

The actual number of channels that can be supported is always less than these idealized figures; the main reasons for this will be described in later sections of this chapter. One of the key factors is the calculation of the RF link budget, taking into account the sources of gain and loss, and its relation to cell size.

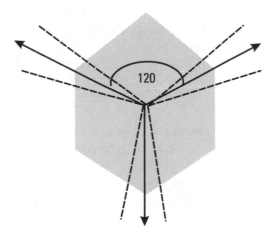

Figure 6.2 Sectors.

6.2   Signal-Quality Factors

Radio propagation is affected by numerous environmental factors that make it much less reliable than communication over a fixed network. An overview of these factors follows.

6.2.1   Path Loss

The most basic of these environmental factors is the loss in signal strength as the signals spread out on the path from receiver to transmitter. With a non-directional (or isotropic) antenna, the signal spreads out as a spherical wave, so the signal strength is inversely proportional to the surface area of a sphere, and hence inversely proportional to the square of the distance from the antenna. In addition to this free-space loss, there are also losses due to other causes such as atmospheric absorption and vegetation. At frequencies of about 2 GHz the only significant atmospheric component is heavy rain, but vegetation is more significant [1]. Leaves on trees are the main culprit, and in wooded rural areas a margin of about 5 to 6 dB has to be allowed for this, while in urban areas it is much less. The free-space loss in the absence of these effects gives an inverse square relationship between transmitted and received powers P_T and P_R, wavelength , and distance d:

$$P_R = P_T c (\lambda / 4d)^2 \qquad\qquad (6.2)$$

where c is a constant.

Allowing for the other effects, the simplest realistic model for path loss is given in logarithmic (decibel) form by

$$P_R = M - 10\beta \log (d/d_0) + e \qquad\qquad (6.3)$$

where M is the measured received signal power at a known reference distance d_0, β is the path loss exponent, and e is a random variable that describes the error between the rest of the formula and the actual path loss. The reference distance is typically of the order of a tenth of the cell radius, while the path loss exponent is the feature that describes losses realistically. It depends on the factors mentioned above and also on the cell size and can be measured empirically, with results usually in the range of 2 to 8, or, as is more usual, standard semiempirical models may be used that contain this factor.

6.2.2 Shadowing

Obstacles that produce shadows will obstruct propagation of the signal in a terrestrial environment. This masking effect is also referred to as slow fading, and is one of the factors that contribute to the error variable *e* in the above path loss formula. The cumulative effects of relatively random obstacles over a wide area are often modeled mathematically by the so-called log-normal law, but over shorter distances specific obstructions may depart significantly from this. The design of the cells has to take account of this to ensure that the probability of a call failing remains acceptably low (e.g., less than 10%) in all conditions. Because the effect is most damaging at the edges of the cells, a shadow fading margin is defined to ensure that the signal strength there remains high enough to meet this probability criterion in the presence of shadowing.

6.2.3 Multipath Interference

Obstacles that are slightly to the side of the direct path will tend to produce reflections, while rows of closely spaced obstacles, such as trees, may also produce diffraction. The reflections have the advantage that signal reception is no longer strictly confined to direct line of sight as some of these may bypass a major obstacle. There are two big disadvantages: the first being a spread in delays, and the second being random frequency modulation due to Doppler shifts resulting from the relative movements of obstacles and antennae. The delay spread causes very rapid fluctuations in signal strength, as signals whose path lengths vary by an odd number of half-wavelengths will cancel each other, whereas those that differ by even numbers will reinforce each other. This effect is also called both Rayleigh fading and fast fading, the latter name resulting from the rapidity of the changes at the short wavelengths used for mobile phones. Together with the Doppler effects, this is the main source of errors (especially error bursts) on the network and limits the signal rate achievable. CDMA has to use fast power control to compensate for the effects of fast fading on slow-moving mobiles, although it is unable to achieve this for full mobility. The increase in power needed at adverse moments means that any design has to include a fast fading margin to allow for this.

In addition to the benefit of multiple path interference in potentially creating an indirect line of sight, modern antennae can combine signals that arrive by different paths by allowing for the different delays with filters and equalizers. Rake receivers [2, 3] perform this signal combination, and their use by WCDMA in both the mobile and the base station is essential. If signals arrive with a separation of more than one chip, then the receiver can resolve them. A rake receiver has several fingers each of which can process a single multipath component by correlating it with a spreading code that is synchronized with the time delay of the path. The components are then weighted according to their strengths and combined. WCDMA-FDD uses either four or nine fingers in the receivers.

6.2.4   Jamming

The signal can be jammed by other transmissions on the same frequency. This consists of random background noise, cochannel interference from other cells using the same frequency, and adjacent channel interference due to harmonics, Doppler shifts, and the slight width of signals on nearby central frequencies. The cell design has to make allowances for this degradation, and does so via an interference margin.

In CDMA the interference due to load causes a significant variation in effective cell size, leading to "breathing in and out" of the cell as the load increases or decreases. The effect is quite large; for example, a cell with effective diameter of 4.6 km when unloaded dropping to 3.6 km at 50%, and down to 0.5 km at 90% [3], so capacity control has to use a maximum load threshold to limit this effect as well as use the interference margin in cell planning.

6.2.5   Handover Gain

A beneficial effect, especially with CDMA, is the effect of handovers, since it can partially counter the adverse effects of shadowing, although this is achieved at the cost of increased complexity. In GSM, handover from one cell to another entails a hard change to another frequency when the difference between current and next cell signal strengths exceeds a set threshold. In CDMA with the same frequency in adjacent cells, this would cause excessive interference in the new cell unless the threshold was very small, which would then cause instability. As a result, CDMA uses soft handovers. In a typical hexagonal macrocell layout each mobile monitors the signals from its six neighboring cells in addition to its own. Because of the different base station locations, the degree of shadowing will vary considerably, and one or more of the neighbors may be much more accessible than the original cell, so that the shadowing can be mitigated by a handover. In a hard handover (usually with frequency change) the mobile switches to the alternative cell that gives the best reception, while in a soft handover (no frequency change and common in CDMA) the mobile always makes use of the base station with the best reception. This leads to a beneficial handover gain that is of the order of 2.5 to 3 dB for hard handovers and 3.5 to 4 dB for soft handovers [4]. A special case of the soft handover is the softer handover that occurs between two sectors of a cell. This differs from the basic soft handover by the reduction in the number of distinct power control commands and in the signaling required between base station components, and it is also faster.

The trade-off between these factors is determined by calculating a radio link budget that can then be used to relate the possible cell size to BTS antenna height using standard design models.

6.3 Radio Link Budget

The link budget determines the maximum path loss that is allowable to achieve the received SNR required for adequate quality. Any antenna has a gain that is formally defined by the expression

$$\text{Gain} = 4\pi \times A_e/\lambda^2 \tag{6.4}$$

where A_e is the effective area or capture area of the antenna and λ is the wavelength of the signal; so large antennae provide more gain. A crude picture of transmission and reception is shown in Figure 6.3 to allow calculation of the power received from a distant transmitter.

In this diagram the powers of transmitter and receiver are $P(T)$ and $P(R)$, respectively, with antennae gains $G(T)$ and $G(R)$, while the cable/connector losses are $L(T)$ and $L(R)$. In addition, there is assumed to be a

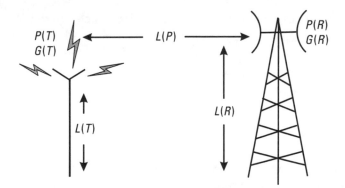

Figure 6.3 Power factors.

total loss on the path between antennae of $L(P)$. The received power is then given in logarithmic decibel format by

$$P(R) = P(T) - L(T) + G(T) - L(P) + G(R) - L(R) \qquad (6.5)$$

where $P(R)$ and $P(T)$ are the logarithms of P_R and P_T in (6.2). These individual factors then have to be further subdivided and related to various margins to work out what is required.

Because of the lower power of the mobile as compared to the base station, the radio budget for the uplink is more critical than that for the downlink in determining the cell size. As a result, in the budget calculations the mobile is assumed to be the transmitter and the BTS the receiver.

The behavior of the BTS receiver is more complicated than that for the mobile, and calculations have to make explicit allowance for noise, interference, and fast fading in addition to the basic parameters above, although these factors also affect the mobile. The fast fading requires a fading margin to be left in the closed power control loop of CDMA that is greatest for slow-moving users.

In addition to the signal, there is also white noise, N, that results from finite temperature of the equipment. This is given by Boltzmann's law

$$N = kTBwF \qquad (6.6)$$

where Bw is the receiver noise bandwidth and F is the noise figure characterizing the receiver. At a typical temperature of 17°C, kT amounts to -174 dBm/Hz, and the noise figure F is typically about $+5$ dBm [5]. For CDMA Bw is the chip rate (e.g., 3.84 Mcps for WCDMA-FDD) and contributes a figure of 10 log (Bw) in decibel units, which is roughly 66 dB for WCDMA, leading to a receiver noise power of about -103 dB in this case. The ratio P_R/N is the SNR that determines the link budget. This is more conveniently expressed as the ratio of the energy per modulated symbol to the noise (E_c/N_0) where

$$(E_c/N_0) = P_R Bw/(NR_c) \qquad (6.7)$$

where E_c is the white noise power spectral density and R_c is the transmission rate of modulated symbols, which is 3.84 Mcps for WCDMA-FDD. In logarithmic form (E_c/N_0) is given by

$$(E_c/N_0) \text{ (dB)} = P(T) + G(T) + G(R) - 10 \log (kT) \text{ (dB/Hz)} - 10 \log R_c$$
$$- 10 \log F - L(R) - L(T) \qquad (6.8)$$

where the factor 10 arises from the use of decibels as opposed to bels.

The interference parameter for the BTS appears as the interference margin to allow for interference from other users and is usually approximated as an additional white noise term. The size of this factor varies according to the loading of the cell by users, ranging from 0 at very light loadings to roughly 3 to 5 dB at 50% to 60%. This combined with the receiver noise density gives the receiver's effective noise plus interference density, which is likely to be of the order of −100 dBm.

CDMA provides another unique type of gain in addition to the antennae gains. This is a processing gain that results from the spreading of the signal over multiple frequencies in accordance with the spreading factor. The degree of this spreading, and hence of the processing gain, depends on the bit rate according to the formula

$$\text{Processing gain} = 10 \times \log \text{ (chip rate/bit rate)} \qquad (6.9)$$

where the bit rate is that of the signal before channel coding and spreading.

In the case of WCDMA-FDD, this becomes $10 \times \log$ (3,840/bit rate in Kbps), and at low bit rates, such as the 12.2 Kbps of AMR voice, amounts to around 25 dB. This allows lower powers to be used for CDMA mobiles for a given bit rate than for earlier GSM varieties.

The receiver sensitivity is given by boosting the noise plus interference density by the processing gain and subtracting the required signal strength. The required signal strength for WCDMA is on the order of 1 to 5 dB with the lower values for data and the higher for voice, while the processing gain varies from about 30 dB at low bit rates down to around 5 dB at peak rates, leading to receiver sensitivities in the rough range of −100 to −125 dBm.

CDMA networks also have the handover gain due to the soft handovers where a mobile is in communication with several different base stations, and so able to compensate for uncorrelated slow-fading and make use of the best signal. This gain is less than the processing gain and usually of the order of 3 to 4 dB.

For the mobile, all that is required is the effective isotropic radiated power (EIRP) given by the previous parameters as

$$\text{EIRP} = P(T) + G(T) - L(T) \qquad (6.10)$$

Typical values for these are on the order of 21 to 24 dB for $P(T)$, zero for $G(T)$, and 3 to 5 dB for $L(T)$, leading to EIRP in the range 16 to 21 dBm.

The maximum acceptable path loss is related to these figures by the following formula:

$$\text{Max path loss} = \text{EIRP} - \text{receiver sensitivity} + G(R) - L(R) - \text{fade margin}$$

(6.11)

A typical value for $G(R)$ is of the order of 18 dB; that for $L(R)$, 2 to 3 dB; and the fade margin 0 to 5 dB depending on the speed of the mobile transmitter. This leads to a maximum path loss potentially in the rough range of 130 to 160 dB.

The actual expected path loss has to be subtracted from this figure in order to determine an allowed propagation loss to use in calculating the cell range for a given antenna height. These factors are the beneficial soft handover gain, a detrimental environmental loss due to the mobile user being in a car or building, the shadow fading margin usually based on the log-normal law, and the fast fading margin. Typical values for these parameters are about 3 dB for the soft handover gain, about 8 dB for being in a car or up to 15 dB for being in a building (as compared to 0 dB for a pedestrian), and a detrimental log-normal factor in the region of 7 dB.

This leads to allowed propagation losses in the rough range of 120 to 150 dB. The actual values vary significantly according to the environment and speed of the user as well as the nature of the service being used, so calculations have to be performed for each category individually.

The allowed propagation loss can then be compared with the loss predicted for a given cell range and antenna height by a standard model. Cell design software packages are usually based on one of two popular models [4]: the Okamuru-Hata model or the Walfish-Ikegama model. The first of these is a purely empirical model based on studies of propagation losses in different types of areas in Tokyo and its surroundings, leading to separate parameter values for suburban and urban areas. The Walfish-Ikegama model is a semi-empirical one in which empirical data is used to give a best-fit value for various theoretical parameters.

One of the outcomes of this type of analysis is that it is best to have different types of cells depending on the type of user and the type of environment. There are three classes of cell size in most networks, as follows:

- *Macrocells:* These have radii in the region of 1 to 30 km and require high mobile power consumption (about 1–10W). They are used for fast movers and rural areas with tall masts. The archetypal layout of macrocells is a grid of hexagonal cells.

- *Microcells:* Characteristic radii are 100 to 300m for use in streets, airports, stations, and other high user-density areas. Antennae are usually below roof level. Two common cell layouts are rectangular, with base stations in the middle of a block, and cruciform, with base stations at a street intersection. The time required to move from edge to center of such a cell at 50 km/hr is only 7 to 22 sec, so it is only suitable for low-mobility users.

- *Picocells:* These are for office environments and use little mobile power (possibly only 10 mW).

There are two limiting design cases for determining the cell layout:

1. *Traffic coverage:* This applies in areas of high traffic density, such as microcells and the vicinity of motorways. The total traffic in the area is divided by the number of channels per cell for a given grade of service (Erlang calculation) to determine the number of cells required. The number of cells then divides the area in order to determine their dimensions, and the propagation loss models used to determine the antenna height needed.

2. *Geographic coverage:* In areas of low traffic density, the cell size is determined by the propagation loss for an acceptable antenna height. This allows the number of cells to be determined by dividing the geographic area by the cell area. The number of channels required per cell is then given by dividing the traffic by the number of cells. These macrocells will frequently contain blind spots that need the addition of repeaters to provide the approximate line-of-sight communication required.

The initial design normally leaves a margin of spare capacity to allow for growth, but eventually additional capacity has to be added.

6.4 Network Expansion

There are a number of techniques for adding capacity with differing scales of expansion, cost, and planning complexity. In the case of 2.5G and third-generation (3G) networks, there is also the desirability of creating the network by developing existing 2G sites rather than starting from scratch, although the additional capacity needed by high data rate services will eventually require extra sites. This has the advantages that much infrastructure is already in place, site survey data needs only minor updates, and planning permission for expansion is relatively easy to obtain. In the case of cdma2000, the changes required are also fairly simple, consisting largely of software and hardware module upgrades, while the use of mobility IP (see Section 7.4.15) minimizes the need for extra gateways. The transition from GSM to GPRS and then to WCDMA entails much more extensive changes in base stations and requires numerous gateways. The gateway issues are covered in Chapter 8 of this book.

The basic means of adding extra capacity and their main characteristics are outlined below:

- *Add extra channels:* This applies particularly to macrocells whose number is determined by geographic coverage. Extra channels can be added by a combination of software reconfiguration and supply of extra transceivers until the maximum for the cell is reached. This is relatively quick, cheap, and free of planning requirements, but is of limited scope.

- *Cell sectorization:* This technique, outlined at the start of the chapter, entails changes to antennae in addition to the transceiver and configuration changes. It adds much more extra capacity but costs more, takes longer, and probably entails additional planning approvals.

- *Extra sites:* This is very slow and expensive, entailing detailed site surveys, redesign of the cell structure, planning approvals, and infrastructure, but also adds considerable capacity.

- *Cell layering:* This is a special case of the addition of extra sites by the addition of picocells within microcells and microcells within macrocells. It is cheaper and more straightforward than addition of the larger cells and makes use of CDMA power control algorithms to minimize interference. In many cases the different layers operate on separate carrier frequencies, so interference is minimized but hard handovers are required. Layering is also required to take account of the different mobility classes of user; for example, it may be practical to use picocells for indoor and pedestrian users, but a motorist on an urban freeway needs to be assigned to a macrocell to avoid excessive handovers.

In UMTS an operator is typically assigned a 15-MHz band for each of the WCDMA-FDD DL and UL, thus allowing the use of three different carrier frequencies. The operator has the freedom to choose how to use these. If macro- and microcells operate on the same frequency, then soft handovers can be used, but the overlap area for handovers has to be a large proportion of the area of the microcell to prevent rapid alternation between layers as measurements of signal strength have to be carried out for a long-enough period to eliminate transient effects of fast fading. This is impractical for all but low-mobility users, so separate frequencies are generally preferred. For micro- and picocells this issue is less critical. The simplest option is one carrier frequency per layer.

6.5   Capacity and Admission Control

In TDMA networks capacity is easily determined, as in (6.1), and is controlled by means of timeslot allocations, but for CDMA, and to a lesser extent GPRS, the situation is more complicated. GPRS for GSM is the simpler of these two and is considered first.

6.5.1   GPRS Admission Control

Timeslots in these networks are shared between CS and PS traffic. The proportion and manner of division of slots between the two functions is up to the operator and may be either static or dynamic; however, in order to guarantee a minimum grade of service for each category, it is likely that a certain number will be reserved for each function, with the remainder in a dynamic pool to be allocated as required. CS allocation is subject to the basic GSM rules of Erlang capacity as outlined in Section 6.1, but PS loads are asymmetric between UL and DL and controlled through the TBFs [6] and the allocation methods described in Section 5.2. UL and DL TBF allocations are performed separately, with multiple TBFs multiplexed into a single PDCH in each case.

6.5.2   CDMA Admission Control

CDMA admission control is based on comparison of the existing load with a threshold value for both UL and DL [4], with a new call accepted only if it can be taken without pushing the cell load over the threshold in either direction. Within a cell there are two types of new call: calls handed over from another cell and brand new calls. Precedence is normally given to the handovers, as it is more irritating to a user to lose a call in progress than to fail to set one up. The most basic method of admission control is to use a channel count, but there are two other more detailed methods of calculating the cell load, of which the simpler is based on power consumption. The load factor LF_{DL} for the DL is particularly simple

$$LF_{DL} = P/P_{Max} \tag{6.12}$$

where P is the total power used and P_{Max} is the maximum available, while that for the UL is slightly more involved. In this case the critical parameter is the received interference power I that is made up of three contributions:

$$I = \text{noise} + I_{Cell} + I_{Intercell} \tag{6.13}$$

where noise is a known parameter of the system, as in (6.6), and I_{Cell} and $I_{Intercell}$ are the variable contributions from the cell and its neighbors, respectively. The UL load factor LF_{UL} is related to these factors via the UL noise rise as

$$\text{UL noise rise} = I/\text{noise} = 1/(1 - LF_{UL}) \tag{6.14}$$

or

$$LF_{UL} = 1 - \text{noise}/I = (\text{noise rise} - 1)/\text{noise rise} \tag{6.15}$$

The base station can measure I and knows the noise, so it can calculate LF_{UL} and compare it with the threshold.

If necessary, congestion control in either direction can be performed primarily through fast power control (FPC). On the DL this is done by rejecting power-up commands from the mobiles; while on the UL it can be achieved by reducing the target SNR that is used by the UL FPC. Slower control options include handing over some calls to an alternative carrier frequency, reducing the traffic rate for flexible algorithms, such as AMR voice, and dropping calls as a last resort.

Estimation of the cell capacity for various different types of traffic requires a more detailed analysis in terms of the load factors for the individual users and their traffic types. The DL load factor based on throughput is given approximately by

$$LF_{DL} = \Sigma_{j=1}^{N} R_j / R_{Max} \tag{6.16}$$

where R_j is the rate of the jth user and summation runs over all N users in the cell, while R_{Max} is the maximum possible aggregate rate and is determined by the chip rate. The power consumption and interference, however, depend also on the SNR for each user, so a more accurate formula that includes these is

$$LF_{\text{DL}} = \Sigma_{j=1}^{N} R_j v_j (E_b/N_0)_j (1 - x_{\text{DL}} + y_{\text{DL}})/W \qquad (6.17)$$

where v_j is the voice activity factor of the jth user, $(E_b/N_0)_j$ is the bit energy measure of the SNR for the jth user and is used in the power control algorithms to determine quality, x_{DL} is the average degree of orthogonality between users in the cell on the DL, y_{DL} is the other cell-to-own cell interference ratio on the DL, and W is the chip rate.

The voice activity factor is a generalization of the factor to describe the inactivity of speech due to half-duplex conversations and pauses. It lies in the range 0 to 1 with typical value (after inclusion of overheads) of about 0.6 for speech and 1 for traditional data applications, with most applications in the conversational class having intermediate fractional values. The average orthogonality and the cell interference ratio both take values in the range 0 to 1, with 0.4 to 0.9 being typical for x and around 0.5 for y. The reason that the orthogonality is less than 1 is mainly due to delay spread resulting from multipath effects and Doppler shifts. $(E_b/N_0)_j$ is the most interesting of the parameters as it decreases with increasing bit rate and has to be increased for a given rate if the frame error rate needs to be reduced for a sensitive application.

An analogous formula also applies to the throughput load factor on the UL. LF_{UL} for the jth user is given by its power consumption P_j as a fraction of the total interference power I, and the total as the sum of these factors

$$LF_{\text{UL}} = \Sigma_{j=1}^{N} P_j / I \qquad (6.18)$$

Expressing this in terms of the bit rates and processing gain leads to

$$LF_{\text{UL}} = \Sigma_{j=1}^{N} \{1 / [1 + W/(R_j v_j (E_b/N_0)_j)]\} \{1 + y_{\text{U}}\} \qquad (6.19)$$

In cdmaOne with predominantly voice traffic, the UL was the more critical direction for capacity on account of the poorer power control in that direction, but for 3G networks this may be outweighed by the much higher bit rates probable on the DL due to downloads from the Web.

Equation (6.17) shows that to a first approximation the number of simultaneous users possible on the DL is inversely proportional to their average bit rates. Use of channel codes (e.g., convolutional or turbo) with rate k/n (see Chapter 2) increases the number of symbols to transmit by a factor n/k and for a constant symbol energy rate also the energy used. In practice it is usual to have a guaranteed energy per bit so that the energy per symbol is reduced by the factor n/k, which leads to an increased symbol error rate at the receiver. The decoded bit error rate, however, is reduced because this is outweighed by the error correction resulting from the information redundancy in the symbol codeword.

The choice of the threshold level depends on the grade of service that is to be provided based on the allowable call blocking probability. This is only determined very roughly by the Erlang B model in CDMA because of the effects of interactions with neighboring cells, notably soft handovers and intercell interference. The number of channels theoretically available for WCDMA-FDD with the parameters quoted in Section 6.3 is of the order of 90 for AMR voice at 12.2 Kbps [4], but allowing for soft handovers reduces this to around 60, and an adequate grade of service gives about 50. For higher bit rate services this drops sharply, but not quite proportionate to the rate as the SIR needed is less at higher rates. By 144 Kbps the maximum is just over 6, but the Erlang formula only allows about 2.5 on average for a suitable grade of service. A side effect of the very low number of channels used is that the cell causes unusually little interference to its neighbors, so that their capacity increases. Consequently, the Erlang principles have to be applied to groups of mutually interfering cells rather than to a single cell. The single cell Erlang capacity is referred to as the capacity with hard blocking, while the modified version is called the capacity with soft blocking. Soft capacity is then defined by

Soft capacity = (hard blocking Erlang capacity/soft blocking Erlang) − 1

$$(6.20)$$

This varies from about 0.05 at speech bit rates to 0.28 at 144 Kbps [4]. Most networks will carry a mixture of call types of which the vast majority are likely to be voice, so the actual soft capacity will be variable and usually much nearer to 0.05. For a cell to make use of the soft capacity, it needs to use a measure of utilization that indicates this, such as the interference power or indirectly by calculation from the number of channels in use and their bit rates.

In WCDMA admission control is handled by the CRNC and Node B (which may support more than one cell) and handled at several stages. The RNC operates admission control based on the interference on the UL and power on the DL for initial UE access, radio access bearer assignment, and handover [7]. The Node B maintains a record of how many resources it is using and passes this information to the CRNC in response to an audit request. A capacity credit scheme is used to estimate how much resource is used on common channels by the NBAP procedures [8]. A cost that is proportional to the SF is subtracted from the capacity credit when a common channel is set up, and added when it is deleted. A similar procedure applies to the dedicated channels. If a single channel uses multiple channelization codes, then the cost of the channel is multiplied by the number of these codes. The Node B also records the DL power in use and the received power and uses a frame priority to decide which frames to transmit during congested periods on the DCH and DSCH.

Many of the 3G data applications are likely to be highly asymmetric, so there is a possibility that the UL will be underutilized in WCDMA-FDD due to the need for a higher proportion of channels on the DL. An operator that has bandwidth for the WCDMA-TDD service in addition to FDD may prefer to allocate as many as possible of the most highly asymmetric applications to TDD as this is able to share the single unpaired band much more efficiently between UL and DL, while putting symmetric applications, such as speech, on FDD.

Cost of Quality

The most direct cost of quality is the reduction in the number of simultaneous users when their bit rates increase. This has to be reflected both in the service charges for different types of calls (e.g., charging by the volume of traffic) and the need to reserve enough capacity for voice calls to ensure an adequate grade of service for them in the presence of transient high demand for data bandwidth. The second main effect is the increase in (E_b/N_0) required for users who need a low error rate. WCDMA-FDD [6] specifies tests for the ratio of the energy per chip (E_c/I_{or}) on the dedicated physical channel (DPCH) to the total spectral density that is required to achieve BLER of 10^{-1}, 10^{-2}, and 10^{-3} at speeds of 12.2, 64, 144, and 384 Kbps on the DL. The tests cover a range of different multipath fading environments, but in general there is a difference of 1 to 2 dB between the low and high BLER.

The changes in strength for the different BLER have a major effect on capacity. A change of 1 dB corresponds to a multiplicative factor equal to the antilog of 0.1, which is about 1.25, while the analogous factor for 2 dB is about 1.6. This leads to changes in capacity of about 25% to 60%, so the network operator's charging policy has to take this into account in addition to the throughput. Many applications include traffic types with differing requirements for error rates (e.g., control data for compressed video as opposed to the image data), and in some cases this may make it worthwhile to use separate channels with different BLER for the distinct types.

The grade of service offered also has a big impact on economics. The load threshold allowed in capacity control has to take account of the Erlang-B probabilities of call rejection. Fixed networks often keep the call rejection rate to 1% or less, while 10% may be more typical on a 2G mobile network. The difference is particularly acute for the case of a small number of high-speed calls. For example, in the instance above of 144-Kbps calls in WCDMA-FDD, changing from a 10% rejection ratio to 1% roughly halves the average load that can be accepted, while for voice calls it would imply a cut of the order of a quarter.

6.6 Power Control

Efficient power control is required to obtain the maximum capacity of a cell. This has to respond to the error rate and the various types of fading as well as the congestion control mentioned above. Outer loop power control responds to the error rate, and in order to do so effectively, it must see a sufficient number of errors in a short time, and this depends critically on the application. If a traffic channel carries radio frames at a rate of 10,000 frames per second for an application that needs a frame error rate of better than 10^{-5}, for example, then it will only see about one frame error every 10 seconds, which will make a reliable response to the error rate impossible in a practical time. On the other hand, if the application can withstand a frame error rate of 10^{-2}, then a sufficient number of errors are likely to be seen within 1 second to allow a reliable adjustment of power within that timescale. In order to support the needs of the high-reliability applications, the solution used in WCDMA and cdma2000 [5, 8] is to monitor the number of bit errors detected (and mostly corrected) by the convolutional or turbo decoder.

Motion of the mobile also requires changes in power. In addition to the overall effect of moving closer or further from the base station, the algorithms have to respond to fading effects. Slow fading due to obstacles takes place relatively slowly and can be handled by slow power control, but fast Rayleigh fading takes place on a very short timescale and so requires the use of fast power control, with WCDMA sending FPC commands at a rate of

1,500 Hz and cdma2000 at 800 Hz. This combination works well for slow fading regardless of the speed of the mobile but is less effective for high or full mobility class users in relation to fast fading. Fast fading takes place over distances of the order of half a wavelength, which at 2 GHz is 5 to 10 cm. During a power cycle of about a millisecond, a pedestrian will only move about 0.1 cm, so FPC can easily handle the fast fading, but a vehicle traveling at 150 km/hr will cover about 4 to 5 cm, thereby rendering the FPC extremely unreliable. The difference in FPC rates between WCDMA and cdma2000 makes the former much better for mobility rates of the order of 100 km/hr.

References

[1]   Tabbane, S., *Handbook of Mobile Radio Networks*, Norwood, MA: Artech House, 2000.

[2]   Steele, R., C.-C. Lee, and P. Gould, *GSM, CdmaOne and 3G Systems*, New York: John Wiley & Sons, 2001.

[3]   Ojanpera, T., and R. Prasad, *Wideband CDMA for Third Generation Mobile Communications*, Norwood, MA: Artech House, 1998.

[4]   Holma, H., and A. Toskale, (eds.), *WCDMA for UMTS*, New York: John Wiley & Sons, 2000.

[5]   Stuber, G. L., *Principles of Mobile Communication*, 2nd ed., Boston, MA: Kluwer Academic Publishers, 2001.

[6]   ETSI TS 125 101, "WCDMA-FDD UE Radio Transmission and Reception," V.4.2.0 (2001), http://www.etsi.org.

[7]   ETSI TS 125 401, "UTRAN Overall Description," V.4.2.0 (2001), http://www.etsi.org.

[8]   ETSI TS 125 433, "NBAP Signaling," V.4.2.0 (2001), http://www.etsi.org.

7

Protocols[1]

7.1 General Remarks

The two network protocols that feature strongly in CDMA and GPRS networks are IP and ATM. CDMA is mandated in both GPRS and cdma2000. UMTS specifies either ATM or IP for the links between RNCs and from RNC to the core network. All these wireless network types, however, are likely to carry IP over ATM in parts of the extended network. Some of the protocols required to meet the needs of 3G networks were still under development at the time of writing in 2001 [e.g., Bearer Independent Call Control (BICC) and Mobility Internet Protocol version 6 (IPv6)]. This chapter describes the features of IP and ATM that are most closely involved with the

---

1. This chapter contains some excerpts from some of the RFCs described above, and these have been used subject to the following RFC copyright notice:

QoS and multimedia applications, both through signaling capability and support for different QoS classes.

7.2 SS7 Features

The UMTS specifications [1–4] anticipate that the core network will make use of the SS7 and broadband signaling protocol [5, 6] to carry traffic over ATM, possibly with IP on top of ATM. Most of the features of SS7 are transparent to QoS considerations, but a minimal sketch of its relationship to ATM and IP is relevant. Figures 7.1 and 7.2 give the outline SS7 protocol stack for the control and user planes.

Figure 7.1 shows the general SS7 architecture for broadband signaling, while Figure 7.2 shows the structure for the specific case of setting up CS speech calls in UMTS between the RNC and core network.

The significance of these component protocols is as follows:

- *Broadband-integrated services user part (B-ISUP):* A user triggers a call by sending a B-ISUP set-up message. At the originating SS7 exchange this sends sets of the initial address message (IAM) to the destination SS7 exchange, which confirms reservation of ATM VPI and VCI resources via an initial address message acknowledgment (IAA) or else rejects it via an IAM reject (IAR) if resources such as bandwidth are not available. The IAM contains the ATM features that are required: bandwidth in each direction, ATM QoS, ATM

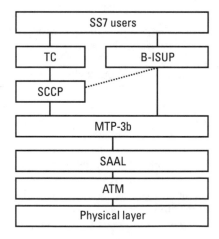

Figure 7.1 SS7 and B-ISUP.

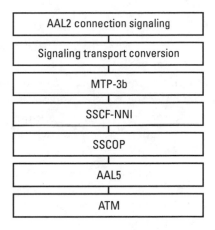

Figure 7.2 Signaling for CS speech.

adaptation layer, and VPI/VCI virtual path and virtual circuit identifiers. If any node on the route is unable to support the required features, the call fails. An answer message (ANM) is sent by the destination exchange to the originating exchange when the called party replies.

- *Transaction capability (TC):* B-ISUP is only responsible for the signaling to set up the ATM virtual circuits; other aspects of SS7 traffic use the TC.

- *Signaling connection control part (SCCP)* [7]: This provides signaling for both connectionless and connection-oriented services between the RNC and 3G-GGSN or MSC, respectively, with separation of the services on a mobile-by-mobile basis. For connection-oriented services it can offer both flow control and error recovery options. It can allow B-ISUP messages to be sent by any route on an SS7 end-to-end basis, but usually this is not implemented and all goes in pass-along mode by the same path as the first signaling message.

- *Message transfer part (MTP-3b)* [8, 9]: This is responsible for routing B-ISUP signaling messages on a hop-by-hop basis, and it can carry up to 4,091 bytes of signaling information instead of the 272 byte limit of the original MTP-3. It reroutes around failures via primary and secondary routes based on predefined link tables, detects errors, and can do congestion control. Where offered, the flow and congestion control is implementation dependent rather than fully standardized.

- *SS7 MTP3-user adaptation layer (M3UA):* This is used instead of MPT-3b when interfacing to an IP core network via the IP Stream Control Transmission Protocol (SCTP) [10] instead of ATM.

- *Signaling ATM adaptation layer (SAAL):* This handles the interface between MPT-3b and the ATM adaptation layer by putting signaling in the right format for ATM-network-network interface signals for the relevant adaptation layer. It has two parts: the service-specific coordination function (SSCF) [11] and the service-specific connection part (SSCOP) [12]. The SSCF lies above the SSCOP and provides functionality that enables the SSCOP to provide data link services for both the user-to-network interface (UNI) and the network-to-network interface (NNI). The SSCOP can support peer-to-peer data transfer in either assured or nonassured modes and checks peer-to-peer connectivity via keep-alives. In assured mode, it provides sequence integrity, error correction by selective retransmission, and flow control.

- *ATM:* ATM adaptation layer AAL5 is recommended for packet-switched data and signaling, whereas AAL2 is preferred for CS speech. Figure 7.2 shows the protocol structure for setting up AAL2 speech circuits by using AAL5 cells for signal transport over the ATM network.

UMTS and cdma2000 also make use of extensions to SS7 to handle mobility administrative functions (such as home and visitor location information transfer) but have incompatible ways of doing so. UMTS uses GSM-MAP, while cdma2000 uses ANSI-41. As a result, gateways have to be provided between the two services to handle data conversion.

7.3 ATM

ATM was originally designed in order to carry both voice and data over the same broadband network in such a way as not to reduce the voice quality compared to that normally provided by Telcos over their pure voice networks, and more generally to support different QoS classes. This objective led to the use of 48-byte cells with additional 5-byte headers and a minimum speed of 51.3 Mbps over an OC-1 circuit—the slowest member of the OC-x SONET optical fiber hierarchy. Because optical fiber circuits have very low error rates (typical BER being on the order of 10^{-10}), ATM makes only minimal provision for error checking, relying on the overlying network application to check and correct errors in the traffic carried and providing a CRC check on the cell header only. An ATM adaptation layer (AAL) is defined to provide a standard way of fitting traffic into the cells in a manner suited to that traffic. There are five such standards (AAL1 to AAL5); the first of which was for constant bit rate traffic such as voice, AAL2 was originally for compressed video (but subsequently redefined), AAL3 and AAL4 were for data compatible with SMDS networks, and AAL5 was for a more bandwidth-efficient way of carrying pure data.

ATM is a connection-oriented service with virtual path and virtual circuit identifiers contained in the 5-byte header. AAL1 carries a 47-byte payload in each of its 48-byte cells, the remaining byte consisting of a 4-bit sequence number and a sequence number check field to correct errors. AAL1 cells are sent at regular intervals over an ATM circuit in such a way as to mimic a traditional voice circuit on a TDM network. Despite requiring more bandwidth than TDM for this purpose, it was quite suitable for uncompressed 64K PCM voice; however, when used for low-bit-rate compressed voice, it faced a major problem with the time taken to fill a 48-byte cell at these low rates (see Table 7.1), as well as with the core network capacity required if partially filled cells were used instead.

There is a delay equal to the cell-fill time at each end of the ATM network that uses up most of the acceptable delay budget for speech at the low bit rates. This delay is increased still further by use of compression algorithms that employ silence suppression. The overall delay of 10 ms for PCM voice is perfectly acceptable, and the core bandwidth figure is an indication of how much bandwidth would be used for other algorithms if partially filled cells were to be used in order to keep the delay to that figure.

Table 7.1
Voice Compression Parameters

Algorithm	Data Rate (Kbps)	Sample Time (μs)	Cell-Fill Time (ms)	Core Bandwidth (Kbps)
PCM	64.0	125	5	72
ADPCM(a)	32.0	125	10	144
ADPCM(b)	24.0	125	13	192
ADPCM(c)	16.0	125	20	288
EVFR	9.6	200	40	580

As a result, many ATM vendors provided proprietary schemes for treating compressed voice as data instead. Eventually, the little-used AAL2 standard for compressed video was modified to become a means of carrying compressed voice without these delays [13]. This is achieved by multiplexing several different traffic streams into a single cell, using the cell and protocol structures shown in Figure 7.3.

There is one service-specific convergence sublayer (SSCS) for each application data stream (e.g., voice call) with a common part sublayer (CPS) responsible for multiplexing them over a single ATM AAL2 connection. The significance of the CPS packet header fields is as follows:

- *Channel identifier (CID):* This 8-bit field identifies the data stream. 0 is not used, 1 indicates layer management, and 2 to 7 are reserved, leaving 248 values for the user traffic.

- *Length indicator (LI):* This 6-bit field gives the number of bytes in the CPS payload, with a default maximum of 44, but with up to 63 possible with splitting between cells. Video data may be more efficient with up to 63 bytes.

- *User-to-user information (UUI):* This 5-bit field carries information such as the type of compression algorithm used for the payload. Value 26 indicates the end of a CPS packet split between two cells, while 27 indicates more of the packet to come.

- *Header error control (HEC):* This is a 5-bit checksum for CID, LI, and UUI;

- *CPS packet payload:* This field contains the application data. The length is variable up to a default length of 44 bytes, but the protocol makes provision for up to 63 bytes.

CPS packet structure SSCS-PDU

CID	LI	UUI	HEC	CPS packet payload

Multiplexing of CPS packets into AAL2 cell

OSF	SN	P	End of last CPS PUD	First SSCS PDU		Last SSCS PDU	PAD

Figure 7.3 AAL2 structure.

Many compressed voice packets will be less than 44 bytes long, so additional headers are required to define their multiplexing into a single AAL2 cell.

- *Offset field (OSF):* This 6-bit field describes the offset into the 47-byte CPS-PDU of the start of the first SSCS-PDU. If the default SSCS-PDU size of 44 bytes is used, then there will be no fragment and the offset will be zero.

- *Sequence number (SN):* This is specified modulo-2, so SN is 1-bit long.

- *Parity (P):* This is a single-bit parity field to protect OSF and SN.

The CPS-PDU payload consists of at least one SSCS-PDU. If the combined length of the SSCS-PDUs is less than 47 bytes, then the field is padded to that length. CPS uses a combined use timer, Timer_CU, to prevent excessive waits before transmitting an AAL2 cell. The value is selected on the basis of the relative importance of low delays or of limiting the core bandwidth used if there are few active data streams. AAL2 requires a special signaling protocol [14] for negotiation of its options; this is ITU-T Q.2630.2. The main control messages for setup are assign request and assign confirm. This includes the CD and optionally the CPS packet specification, one of about eight possible SSCS selections, and the UUI.

Voice on mobile networks uses any one of a variety of low-bit-rate codecs (e.g., 12.2-Kbps EFR or AMR on UMTS and 14.4 Kbps or 9.6 Kbps on cdma2000, as well as subrates of these). As a result, AAL1 is unsuitable, and mobile voice traffic over ATM is carried in AAL2 cells. Data for mobile networks uses the efficient AAL5 cells in preference to AAL3/4 that have 4-byte internal headers in each of the 48 bytes, and radio signaling traffic is also carried in AAL5. Although AAL2 can support packet fragmentation and thus carry video, it is not recommended for this with standards like H.323 where integrity will be lost [15].

ATM supports QoS. Originally four QoS classes—labeled A, B, C, and D—were defined by the ITU in I.356, but these were not very helpful and have been replaced [16] by a set of ATM transfer capabilities (ATC) that are roughly aligned with an alternative set of service categories defined by the ATM Forum. These are tabulated in Table 7.2.

The B-ISUP signaling in SS7 supports constant bit rate (CBR), variable bit rate (VBR), ATM block transfer (ABT), and available bit rate (ABR) [but not unspecified bit rate (UBR)]. The VBR category is characterized by a sustainable cell rate (SCR) and is split into three subcategories: (1) VBR.1

Table 7.2
ATM QoS Categories

ATCs	QoS Class
CBR, VBR.1, ABT	Class 1 (stringent)
CBR, VBR.1, ABT	Class 2 (tolerant)
VBR.2, VBR.3, ABR	Class 3 (bilevel)
Any ATC	Class U (unspecified)

specifies cell delay time (CDT) and cell delay variation (CDV), (2) VBR.2 excludes CDV, and (3) VBR.3 merely uses the cell loss priority (CLP) bit to control cell dropping if the SCR is exceeded. Stringent and tolerant QoS set limits on cell loss probability regardless of CLP (to 5×10^{-7} and 10^{-5}, respectively), while bilevel applies the loss to cells with CLP = 0 only with a limit of 10^{-5}. Stringent also limits the CDT, while Class U sets no limits.

ABT replaces an earlier standard called real-time VBR and enables data to be sent as blocks that behave like the CBR standard, but for short periods of time with block cell rates (BCR) that differ from one block to another. It also uses the SCR, and the BCR is only allowed to exceed the SCR until an agreed expiry time and must always be less than a peak cell rate (PCR). In order for the BCR to be changed, there has to be an exchange of resource management (RM) cells. These RM cells were originally introduced for the ABR service (see below), and although their format remains the same, the interpretation of the fields is different. The key fields are a requested new BCR in an ABT RM request and a CI flag in the corresponding acknowledgment that indicates whether the new BCR has been accepted or not. ABT is arguably the most suitable ATC for compressed video applications, such as MPEG-2 (see Section 10.2.4).

ABR uses switched virtual circuits within whatever is the available bandwidth. This is in contrast to the other types that depend on permanent virtual circuits. ABR needs detailed control and makes extensive use of RM cells that carry such features as a congestion indicator (CI) to force a rate reduction, and a no increase flag, which, if set, would prevent increase of the cell rate by a preagreed amount that would otherwise be permitted.

In the ATM UNI v4.0 specification [10], the ATM Forum introduced some extended QoS parameters, including desired and cumulative CDV and acceptable cell loss ratio in each direction, but these are not supported by the SS7 B-ISUP standards applicable to UMTS.

7.4   IP

IP is the protocol with the greatest influence on the QoS as it is used end-to-end and transported only within ATM over restricted parts of the network. Features of IP are defined in the Requests for Comment (RFC) from the Internet Engineering Task Force (IETF). The original form of IP is IP version 4 (IPv4), which uses 4-byte addresses and initially assumed that all end stations were fixed devices. As its usage has increased, it has been adapted to meet a wider range of requirements, particularly mobile hosts and longer addresses to cope with the increased number of devices on the Internet. In addition, the increasing range of applications being carried over IP has necessitated modifications to give better support for QoS and has also led to the development of specific subprotocols, such as RTP and ESP. This section discusses Ipv4 and IPv6, Mobility IP, QoS features, and the main subprotocols. The use of IP in UMTS and cdma2000 is covered in Chapters 8 and 9.

The UMTS and cdma2000 standards bodies—3GPP and 3GPP-2, respectively—have agreements with the IETF for cooperation in the use and development of RFCs.

7.4.1   IPv4

The reader is assumed to be familiar with IPv4, so this section gives just a brief resume of relevant aspects plus a description of some of the main shortcomings of its use in 3G networks.

From its inception IP contained an 8-bit type of service (TOS) field that could in principle be used for QoS (see Section 7.4.5). In practice, however, little use was made of this facility, and most networks instead made use of proprietary prioritization schemes in routers based on filters for source and destination addresses, protocols, and applications. This process was effective for data applications, but was labor intensive to set up and maintain and barely adequate for voice and multimedia. Recently, the IP precedence subfield consisting of the first 3 bits of the TOS have been widely used to help prioritize packetized voice applications, such as VoIP, but much more has to be done for this on low-speed links. On such links the basic problems are delay and jitter resulting from queuing behind large data packets, as even a single 1,500-byte packet on a 128-Kbps link would cause a delay of about 90 ms, thereby using up almost the entire delay budget for acceptable voice communication (see Chapter 9). The main cures for this are the segmentation of all packets, at least to the minimum MTU size of 256 bytes, and the possible use of resource reservation via the Resource Reservation Protocol

(RSVP) for integrated services (see Section 7.4.3). Where required, dynamic addressing is covered by DHCP [17].

7.4.2   IPv6

The main rationale for the introduction of IPv6 [18] was the lack of a sufficient number of IP addresses in IPv4. This led to the replacement of 4-byte addresses by 16-byte addresses. The initial shortage of addresses was largely circumvented for fixed networks by the use of private internal corporate network numbering based on the use of the Class A network 10.0.0.0 and the Class B networks 172. 16-31.0.0, together with network address translation for external communication via a firewall on an official Class C network. This approach has the advantage of additional security but is unsuitable for isolated mobiles that are likely to be both semipermanently on and mobile. The simplest option is for these devices to have their own permanent IP addresses, which will rapidly force the use of the much more numerous IPv6 addresses. Other innovations, such as home networks for intelligent home equipment, will add to the demand for more permanent IP addresses and, hence, the use of IPv6.

A second reason for the preference of IPv6 to IPv4 is the fact that the use of the TOS field is mandatory instead of optional as in the earlier version. IPv6 also supports larger packet sizes and hop limits (65,535 octets and 255 hops) than IPv4, reflecting the requirements of faster low-error networks. Another important feature is that IPSec is a standard feature of IPv6 instead of an optional add-on for security in IPv4. These main changes necessitate a different header structure for IPv6, and this is shown in Figure 7.4.

The traffic class is an 8-bit field that plays essentially the same role in IPv6 as TOS does in IPv4 (see Section 7.4.5). The default setting has each bit set to zero, but early definitions [18] do not specify the nontrivial settings.

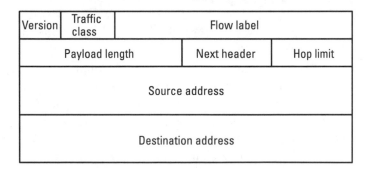

Figure 7.4 IPv6 header.

A useful new feature, potentially applicable to QoS considerations, is the 20-bit flow label. The main purpose of this label is to provide a quick way for an IPv6 router to recognize packets that belong to a sequence that requires some nondefault QoS treatment. Hosts or routers that do not support this function are required to set it to zero if originating the packet, but otherwise they are to leave it untouched. Details of the positive use of the label are not defined in the basic standard [18].

The 16-bit payload length gives the length of the payload in octets including any of the optional extension headers in that sum. Next, header is an 8-bit field that describes the next header after the main header in terms of its protocol as per RFC1700. Hop-limit is also an 8-bit field that can be set as high as 255 (instead of 15 for IPv4) for very large networks and is decremented by one for each hop. The 128-bit source and destination IP addresses complete the main header. Because some of the old IPv4 fields have been dropped or made optional, this header is shorter than the previous one despite the increase in address length.

Optional extension headers may be inserted between the main header and individual upper-layer (e.g., UDP or TCP) headers. The next header field of the previous header identifies each of these extension headers. Only one of these headers is examined and used en route until the final destination; this exception is the hop-by-hop options header, which if present must follow the main header directly. The other possible extension headers listed in their preferred order of appearance is as follows: destination options, routing, fragment, authentication, encapsulating security payload, destination options (2), and upper-layer header. The first of the destination options headers is for options that must be processed by the destination address in the main header and also by any addresses in the routing header, if present, while the second is for options that only the final destination processes.

In relation to QoS the most relevant of the optional extensions is the fragment header, identified by next header value 44. Fragmentation is handled differently in IPv6 as compared to IPv4. Instead of being performed by any router along the path from source to destination (as in IPv4), it is only carried out by the source node in IPv6. The packet consists of an unfragmentable part plus a fragmentable section. The unfragmentable part consists of the standard IPv6 header together with any extension headers that have to be processed en route by routers or other nodes. The structure of the fragmentation header is shown in Figure 7.5.

Next header is an 8-bit field that defines the initial header of the fragmentable part of the packet. Initially, the 8-bit reserved field is set to zero. The fragment offset is a 13-bit unsigned integer that gives the offset in

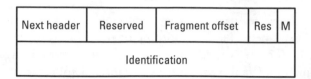

Next header	Reserved	Fragment offset	Res	M
Identification				

Figure 7.5 IPv6 fragmentation.

8-octet units to the data following this header, relative to the start of the fragmentable part of the original packet. Following another 2-bit reserved field initially set to zero is the M-flag to indicate more fragments to follow and the 32-bit identification field that gives a numerical identifier to the packet being fragmented. This is usually treated as a wrap-around counter incremented by one for each packet to be fragmented either on a per-address basis or in total depending on implementation. The destination host reassembles the fragments as usual. The other difference in fragmentation is the minimum maximum transmission unit (MTU) size of 1,280 bytes in IPv6, with a strong recommendation for use of at least 1,500. The packet should be reassembled within 60 seconds or else an error message is sent, likewise if the final packet exceeds 65,535 octets.

IPv6 also uses a modified version of the Internet Control Message Protocol (ICMP). Dynamic addressing is different than in IPv4 and is handled according to RFC2462 [19].

7.4.3   Integrated Services

The central idea behind integrated services [20] is to provide an application with the ability to choose its required QoS from a range of controlled options provided by the network. This in turn depends on the nodes of the network having the ability both to control the QoS and to have a means of signaling the requirements. In principle, it is possible to use manual configuration or Simple Network Management Protocol (SNMP) commands to set the QoS requirements at each node, but this is only practical for tiny networks, so a proper signaling protocol is required, and this is provided by RSVP. There are also two basic control options for integrated services: controlled load [21] and guaranteed delay [22]. Controlled load is the simpler of the two and is intended for the many real-time applications that operate well at light loads but degrade badly in the presence of network congestion. Its aim is to ensure that these applications do not encounter congestion, and it tries to achieve this by admission control. Guaranteed delay is intended for applications with strict end-to-end delay requirements. It calculates a worst-case delay that it is able to guarantee, but does not specify either an average delay or any limit on jitter.

These control options make use of traffic specifications by both the sources and receivers and a feedback mechanism to show what is being provided. In RSVP, the receiver is responsible for stating its flow requirements in a FLOW_SPEC that is sent back upstream to the sender, and that can be modified by intermediate nodes. This FLOW_SPEC consists of two parts: a TSPEC that characterizes the traffic and, where required, an RSPEC. Similarly, the sender generates a SENDER_TSPEC that characterizes its traffic, and this is sent unchanged through the network to the receiver. These traffic specifications are both based on a TOKEN-BUCKET_TSPEC that contains five 32-bit fields as follows:

1. The token-bucket rate (r) given as the number of bytes of IP datagram per second, with values in the range of 1 byte up to 40 TB/s;

2. The bucket size/depth (b) given as a number of bytes from 1 byte up to 250 GB;

3. The peak rate (p);

4. The minimum packet size (m) generated by the application including all IP headers (such as UDP, TCP, and RTP) but not link-level headers;

5. The maximum packet size (M), which should be the minimum acceptable MTU on the path between sender and receiver.

The feedback message is the ADSPEC, which also contains 32-bit fields that define a summary of the available service characteristics. It is updated as it passes back through each node from sender to receiver. It always includes four general fields, and additionally has sections specific to the controlled load and guaranteed delay services as appropriate. The four general fields are as follows:

1. HOP-IS-COUNT;

2. Path Band-Width Estimate;

3. Minimum Path Latency;

4. Composed MTU.

There are also flags to indicate whether any nodes on the path fail to support RSVP.

For the guaranteed delay, there is information on two parameters, usually called C and D, which are explained in the paragraph below on that service. The controlled load service can operate without the ADSPEC and no extra information is usually passed for it.

The HOP-IS-COUNT indicates the number of IS-aware-of-RSVP nodes along the path and is incremented by one by each such node. The flag for one or more unaware nodes on the path may have to be set manually, since such a node will not know how to set the flag.

The path band-width is the minimum of the estimates by the individual nodes and is composed by each node either copying the previous entry or inserting its own estimate if that is lower. The parameter is quoted in bytes per second and is likely to be either the bandwidth of the relevant outgoing link or a fraction of that depending on other traffic.

The minimum path latency is an estimate of the fastest possible network transit time in microseconds and includes all fixed contributions such as propagation delay and minimum node processing time, but excludes all queuing delays. Where multiple paths exist between source and destination, and also for multicasts, this parameter is potentially ambiguous, and if a single value only is given, it will be that for the quickest path. This means that the total transit time for a guaranteed service packet will normally be greater than the sum of the guaranteed delay and the minimum latency.

The composed MTU is required because MTU discovery works from the source, not the receiver, and an application may also need to have a maximum MTU that is less than the configured MTU for a hop. The value computed at any node is the lesser of the previous composed MTU and the node's own maximum MTU.

Controlled Load Service

This service provides its users with a QoS comparable to a best-effort endeavor on a lightly loaded network, but at higher loads. This means that there should be little or no queuing over timescales that exceed the bucket size divided by the bucket rate. In order to achieve this, it has to be able to provide a temporary flow rate exceeding the bucket rate to clear any backlog immediately after a traffic burst. It also needs to try to avoid dropping packets for this service, and to achieve this aim it needs buffering that exceeds the maximum burst size. Traffic policing is required to ensure conformance, and the criterion used is that over any time period T, the quantity of data sent should not exceed $(rT + b)$, where r and b are the token bucket parameters from the TSPEC. An additional characteristic is that links must not segment packets conforming to the service, but service packets larger than the MTU are classified as nonconformant.

Guaranteed Delay Service

This service guarantees that the delay will not exceed a specified maximum and also that no packets belonging to the service will be dropped due to congestion if the flow remains within specification. The aim is that the traffic flow for this service should look as similar as possible to a fluid flow based on a bandwidth of R and a buffer size of B. Two parameters, C and D, are calculated for the path that characterizes the maximum deviation from this ideal; C (measured in bytes) is rate dependent, while D is rate independent. The queuing delay for any datagram must be less than $(b/R + C/R + D)$. The FLOW_SPEC for this service contains a TSPEC as before, but also an RSPEC. The RSPEC consists of two terms: a rate (R) that satisfies $R \geq r$, and a slack parameter S. The slack, S, determines the difference between the required delay and the delay that would be achieved at the rate R in the RSPEC.

The guaranteed maximum queue delay in terms of these parameters is as follows:

$$\text{Max delay} = (b - M)/[R(p - R)/(p - r)] + (M + C_{total})/R + D_{total}$$
$$\text{for } p > R \geq r \tag{7.1}$$

$$\text{Max delay} = (M + C_{total})/R + D_{total} \text{ for } r \leq p \leq R \tag{7.2}$$

where C_{total} and D_{total} are the cumulative totals for C and D over all nodes in the path. C represents the delay that a datagram might experience due to the rate parameters of the flow, such as the serialization delay for a datagram broken into ATM cells. D is the delay on top of the minimum latency at a node due to waiting for a slot in a slotted service or for the transmission of a worst-case MTU on the output link of a router just before the guaranteed datagram. The slack, S, is related to C and D by the following:

$$S = \text{Delay\_Required} - (b/r + C_{total}/r + D_{total}) \tag{7.3}$$

Positive nonzero S can be used to allow acceptance of other flows that would not be acceptable otherwise.

The traffic in the guaranteed service is policed at the edge of the network, and the data sent in any time interval T must not exceed a value given by $(M + \text{minimum}[pT, (rT + b - M)])$.

The additional terms in the ADSPEC for a guaranteed service are C_{total} and D_{total}, together with the cumulative values of C and D from the last reshaping point (i.e., at the last heterogeneous branch point or the last source merge point).

7.4.4 Resource Reservation Protocol

RSVP [23, 24] is a unidirectional control protocol that enables the QoS to be signaled and controlled; this is an option for QoS control in UMTS (see Sections 8.1.3 and 9.2.3). The receiver is responsible for requesting the QoS backwards until reaching either the sender or a multipoint distribution point. Each node along this path negotiates with its own link level to either accept or reject the reservation request; all must accept for establishment of an RSVP session. Nodes that do not support RSVP pass the messages transparently and do not count as rejections. An RSVP session is characterized by a destination address, protocol ID, and a destination port number (in TCP or UDP or potentially other equivalent). An RSVP resource reservation request consists of the FLOW_SPEC, as described above, and a FILTER_SPEC, which defines which set of packets is to receive the QoS specified. RSVP is supported in both IPv4 and IPv6, but the FILTER_SPECs differ according to this type.

The use of UDP/TCP port numbers in the session definition prevents IP fragmentation, as this information is not carried in each fragment, and this is the reason for not allowing this in the two integrated service types above. Encryption can also hide this information, but RFC2207 [25] provides a way around this problem. Variable size headers in IPv6 also make detection of these port numbers harder, and use of the IPv6 flow label field is a potentially easier alternative in that case.

RSVP contains two classes of messages: reservations (Resv) and path. The first class contains reservation requests, errors, and confirmations. Path messages originate from the sender and follow the path back to the receiver that the unicast or multicast traffic will follow. One of the most important path messages is the ADSPEC.

RSVP messages have a standard set of formats, all of which include a common header with the structure shown in Figure 7.6.

Version	Flags	Message type	RSVP checksum
Send_TTL		Reserved	RSVP length

Figure 7.6 RSVP common header.

The 4-bit version field is the RSVP version, currently version 1. There are seven possible message types as follows: 1 = Path, 2 = Resv, 3 = Path Err, 4 = Resv Err, 5 = Path Tear, 6 = Resv Tear, and 7 = Resv Conf.

The 16-bit RSVP checksum covers the whole message and uses one's complement.

The Send_TTL is the IP TTL for the message, while the RSVP length in bytes is the total RSVP message length including this header.

The common header is followed by one or more object formats that consist of one or more 32-bit words together with a one-word object header. This header consists of a 2-byte object length field followed by a 1-byte class number and 1-byte client type (c-type). Each object has a name and a class number, while the c-type is mainly used to distinguish between IPv4 and IPv6 of an object with 1 for IPv4 and 2 for IPv6. Table 7.3 lists the main object types.

RSVP FLOW_SPECS can be merged as they back upstream from receivers to senders. TSPEC merging is performed for controlled loads by taking the largest bucket rate (r), the largest bucket size (b), the largest peak rate (p), the smallest minimum packet (m), and the smallest acceptable MTU (M).

RSVP has a significant synergy with ATM as regards QoS. Both treat source-destination flows on an individual basis (apart from the merging) and both use comparable flow specifications. In particular, the RSVP bucket rate is analogous to the SCR in ATM, while the peak rate is analogous to the PCR.

The use of RSVP and integrated services does not solve all QoS issues, particularly where slow links are involved. In the example of the 1,500-byte packet on a 128-Kbps link mentioned earlier in this section, the use of a minimum MTU for IPv4 would cut the packet to 256 bytes leading to a worst-case queue delay (or D component for the guaranteed delay service) of 16 ms, which is still higher than desirable. In the case of IPv6 the minimum

Table 7.3
RSVP Objects

Number	C-Type	Name/significance
0	—	Null
1	1, 2	Session. IP dest addr, IP protocol ID, Dest Port
3	1, 2	RSVP_HOP. IP addr of msg sending node plus O/G I/F
4	1	Integrity. Cryptographic data for authentication
5	1	Time-Vals. Refresh period for RSVP RESV
6	1, 2	ERROR_SPEC. For PathErr, ResvErr, or ResvConf
7	1, 2	SCOPE. IP addresses of sender hosts addressed by msg
8		FLOW_SPEC. See above sections
9		STYLE. Reservation style
10	1, 2	FILTER_SPEC. Source IP address and port
11	1, 2, 3	SENDER_TEMPLATE. C-type 3 uses IPv6 flow label
12		SENDER_TSPEC. See above sections
13		ADSPEC. See above sections
14		POLICY_DATA. Policy defined by common open policy server (COPS)
15	1, 2	RESV_CONFIRM. IP add of receiver req confirmation

MTU is 1,280 bytes, so the corresponding D component is an unacceptable 80 ms. Both integrated services and IPv6 in general have to make use of sub-MTU segmentation at the link level to overcome this problem for delay-critical applications such as voice. An additional issue that is not addressed by the integrated-services approach is the variability of delay, or jitter. The delay for a guaranteed delay service will vary from zero up to the guaranteed maximum, and will often exceed acceptable levels. The basic cure for this is to provide a playback buffer on the end devices for real-time applications to enable smoothing out of this variation at the expense of increasing the average delay towards the guaranteed figure. The size of the playback buffer is determined by the guaranteed delay.

A further issue with the use of RSVP is that apart from some specific features such as the ADSPEC, the protocol is not very scalable, making it unsuited to use in large networks. On a very large network there are likely to be many users on trunk routes with similar QoS requirements, so it is much more efficient to have a collective approach to handling their needs. This is performed by the differentiated services protocol described in the following section.

Extensions to RSVP have been defined for MPLS ([26] and Section 7.4.13).

7.4.5 Differentiated Services

Differentiated services (DiffServ) [27] may be used with both IPv4 and IPv6. They are based on use of a standard differentiated services interpretation of the TOS field in IPv4 and the traffic class field in IPv6. The purpose is to provide a scalable approach to QoS that does not require detailed per-hop signaling. The differentiated services architecture contains two main components: (1) the use of simple queue priority mechanisms for forwarding, and (2) the use of allocation policies and configuration rules at the nodes.

The key feature of the forwarding scheme is the reinterpretation of the TOS, or traffic class, as the DS field. This remains an 8-bit field, but the first 6 bits become the DS code-point (DSCP) field, while the last 2 bits are currently unused (CU) and ignored. DSCP is always treated by a DS-compliant node as an unstructured 6-bit field that indexes a table of per-hop-behavior (PHB). Thus, a DS-compliant node lumps all packets with equal DSCPs into a single class or behavior aggregate and selects the appropriate PHB from its table. This table must always contain a default PHB that is to be used for any packets that carry an unrecognized code-point. This usage is incompatible with the original TOS field, so the network needs to be divided into DS domains that support DS functionality and other regions that do not, as well as to have boundary nodes that translate between the two types of usage. Recommended PHBs are compatible with the interpretation of bits of the TOS as IP precedence (bits 0–2) [28], but not with the original TOS specification (RFC791). These compatible DSCPs have the form $xxx000$ and are called class selector code-points. They select from at least two and up to eight different traffic-forwarding classes. A class selector code-point with a larger numerical value than another is said to have a higher relative value and should give a PHB forwarding probability that is not lower than the other. DSCP 000000 is the default and gives the lowest priority of forwarding. The possible 64 different code-points are divided into three pools, of which pool 1 defined by values $xxxxx0$ is the standard set, while pools 2 and 3 are the 16-element sets $xxxx11$ and $xxxx01$, both of which are reserved for experimental or local use. The precise mechanism of the PHBs was not specified in the original RFC [27] but was added later [29, 30].

Assured forwarding (AF) is a type of PHB introduced [29] to allow different drop precedences for packets within a quality class in the event of congestion; this has some resemblance to Intserv controlled load. Four AF classes are defined, and within each there are three levels of drop precedence. Packets belonging to different AF classes are treated independently, and such classes are never aggregated. A DS node has to reserve a minimum number of resources for each AF class in use and has to ensure that the probability of forwarding an IP packet with one given precedence value must be no smaller than that for forwarding a packet with higher drop precedence in the same AF class. It allows the number of precedence levels to be effectively reduced to two in relatively uncongested networks. Packets are not reordered within the same AF microflow.

The precise dropping algorithm is not specified, but should respond gradually to increasing congestion and use queuing to handle transient bursts and dropping to handle longer-term congestion. In practice, both the dropping and forwarding algorithms depend on proprietary features within the networking equipment. The DS code-points for these AF classes are indicated in Table 7.4.

Another form of PHB, called expedited forwarding (EF), is defined in RFC2598 [30] and is also referred to as premium service. The purpose of EF is to provide a virtual point-to-point pipe for traffic that needs a low delay and minimal jitter, but it is only an optional extension to DiffServs.

EF depends on guaranteeing a minimum departure rate from a DS node for the EF flow regardless of the level of other traffic at that node. The implementation mechanism is not specified, but requires a form of priority queuing or bandwidth reservation. The recommended DS code-point for EF is 101110.

Table 7.4
DSCP for AF Classes

Class	AF1x	AF2x	AF3x	AF4x
Drop precedence				
Low	001010	010010	011010	100010
Medium	001100	010100	011100	100100
High	001110	010110	011110	100110

Note: x is 1, 2, or 3 for low, medium, or high, respectively.

Differentiated services can be used with RSVP in the core of a network to reduce the scalability problems of that protocol [31]. Effective use of Diff-Serv in a large network, such as a 3G mobile network interacting with fixed users, depends on having suitable service level agreements (SLA) between the constituent subnetwork operators, and the use of thorough traffic engineering. The SLAs are implemented at the DS-egress node from one domain to the DS-ingress node of another using some form of traffic conditioning. In general, this entails measuring the amount of traffic in each class (behavior aggregate) and mapping it onto those of the other network. Frequently this will involve rewriting the DSCPs as there is no guarantee that both networks will support the full range. PHBs may easily be restricted to the EF premium service, one or two AF classes, and a default. Traffic policing has to be performed in such a way that an agreed quota for a certain class of traffic through the DiffServ core is used to provide the required QoS to a limited number of users rather than an inadequate service to a larger number. (See [30] for a detailed discussion of the issues and solutions.)

7.4.6 Common Open Policy Server

Many networks use some form of policy server to help to control QoS, particularly admission control. Common open policy server (COPS) [32] is a standard for a client-server query/response protocol for this purpose. The server is called a policy decision point (PDP) and the clients are called policy enforcement points (PEP). Section 8.1 includes examples of the use of COPS in UMTS networks. COPS provides reliable messaging through the use of TCP for transport control and can provide authentication security through either IPSec or TLS.

The PEP is responsible for setting up the TCP connection to the PDP and uses it to request and receive decisions from the remote PDP, or occasionally from a local PDP (LPDP) in a hierarchical arrangement, where the decision information is relayed from PDP to LPDP. Fault tolerance is achieved through the use of keep-alive messages between PDP and PEPs, and a timer to allow the PEP to continue using the last PDP decision for a limited period while trying to reestablish a failed PDP-PEP connection. One PDP implementation per server must listen for these TCP connection requests from PEPS on TCP port number 3288.

COPS messages use a common header format as shown in Figure 7.7.

The 4-bit version refers to the COPS revision level, and the initial version is 1.

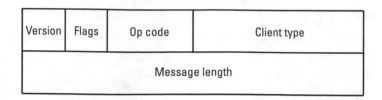

Figure 7.7 COPS common header.

There is currently only one flag defined in this field, the solicited message flag, which indicates that another COPS server has solicited the COPS message.

The 8-bit op code indicates the type of COPS message by means of the values in Table 7.5.

The c-type is a 16-bit field whose value determines the interpretation of the encapsulated objects that follow the common header. The 32-bit length field gives the length in octets of the COPS message including the common header. The encapsulated objects each consist of one or more 32-bit words (padded if necessary to make integral words), with the first word of each object being an object header as in Figure 7.8.

The c-num field describes the type of object according to Table 7.6.

Table 7.5
COPS Op Codes

Op Code	COPS Operation
1	Request (REQ)
2	Decision (DEC)
3	Report state (RPT)
4	Delete request state (DRQ)
5	Synchronize state req (SSQ)
6	Client-open (OPN)
7	Client-accept (CAT)
8	Client-close (CC)
9	Keep-alive (KA)
10	Synchronize complete (SSC)

Figure 7.8 COPS object header.

Table 7.6
COPS Object Types

C-Num	Nature of Object
1	Handle
2	Context
3	In interface
4	Out interface
5	Reason code
6	Decision
7	LPDP decision
8	Error
9	Client-specific information
10	Keep-alive timer
11	PEP identification
12	Report type
13	PDP redirect address
14	Last PDP address
15	Accounting timer
16	Message integrity

The c-type distinguishes between different types or versions of the c-num. A frequent distinction is the use of c-type 1 for IPv4 and c-type 2 for IPv6, but the decisions have several separate c-types.

The client handle is used in most COPS operations and is selected by the PEP to identify the request state of a particular client type. It is then used in requests, reports, and delete messages from PEP to PDP until deleted (for the reason specified in the reason code of a DRQ).

The context object is required for request messages and specifies the type of event that triggered the request. The structure of this object is shown in Figure 7.9.

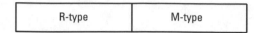

Figure 7.9 Context object.

Request type (r-type) flag

- $0x01$ = Incoming-Message/Admission Control request
- $0x02$ = Resource-Allocation request
- $0x04$ = Outgoing-Message request
- $0x08$ = Configuration request

Message type (m-type)

- Client-specific 16-bit values of protocol m-types
- The key policy control objects are the decisions (c-nums 6 and 7) issued by the PDP or LPDP, respectively. There are three possible outcomes:

  1. Command-code 0 indicates a NULL decision (no configuration data);
  2. Command-code 1 indicates INSTALL (accept request/install configuration);
  3. Command-code 2 indicates REMOVE (remove request/remove configuration).

There are at least five c-types describing different types of decision data.

Use of COPS with RSVP

COPS does not require the use of RSVP but can be combined with it through the use of the policy objects for RSVP as defined in an extension to the RSVP definition [24]. RSVP is potentially only one of many c-types and corresponds to COPS c-type 1. COPS provides admission control, while RSVP provides resource allocation for connection requests that have been accepted by COPS.

Policy data is carried in the type 1 RSVP class 14 Policy_Data Object for c-type 1. After the RSVP object header there is a policy options list and a set of policy elements, both of which are opaque to RSVP but used by policy-aware nodes along the route to ensure consistent policies. RFC2753 [33] discusses principles for implementing PDPs and PEPs in conjunction with RSVP routers or in general, and these recommendations are followed in UMTS (see Section 8.1.3).

7.4.7 Real-Time Transport Protocol

Real-Time Transfer Protocol (RTP) [34] was designed to provide end-to-end network transport functions suitable for real-time data transmission, especially audio and video, over both unicast and multicast network services. Unlike RSVP it does not reserve resources, nor does it guarantee QoS. Applications usually run RTP over UDP in order to use the multiplexing and checksum functions of the latter. Examples of applications likely to be used over 3G networks that will run over RTP are given in Chapter 10. The features that RTP provides include payload identification, sequence numbering, time stamping, and delivery monitoring. It does not reorder out-of-sequence packets nor provide timely delivery, but the sequence numbers and time-stamps enable the end application to perform this itself. RTP uses its own Real-Time Transport Control Protocol (RTCP) to provide feedback on the quality of data distribution, carry a persistent name for the RTP source, and perform some multicast session control. Multimedia applications, such as video-conferencing, use separate RTP sessions for each media type.

RTP adds a header to its payloads that has the structure shown in Figure 7.10 where

1. Version (V) is a 2-bit field defining the version of RTP. That for RFC1889 is 2.
2. Padding bit (P) is set if one or more padding octets are added at the end of the payload where fixed block sizes are required for another protocol.

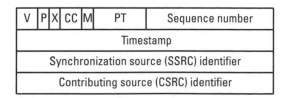

Figure 7.10 RTP header.

3. Extension bit (E) is set if there is exactly one header extension after the fixed RTP header.

4. CSRC count (CC) is a 4-bit count of the number of CSRC identifiers that follow the fixed header.

5. Marker bit (M) can be used to mark important events, such as video frame boundaries, in the data stream.

6. Payload type (PT) is a 7-bit field to identify the type of payload by a standard code.

7. Sequence number is a 16-bit identifier for individual RTP packets that is incremented by one for each RTP data packet sent.

8. Time-stamp is a 32-bit time that indicates the sampling time of the first octet of the data packet. Packets that belong to the same video frame typically have the same time-stamp.

9. Synchronization source (SSRC) identifier is a 32-bit field that is chosen at random to prevent any two synchronization sources within a single RTP session from having the same value.

10. Contributing source (CSRC) identifiers is a list of up to 15 contributing payload sources for this packet, each being identified by a 32-bit entry.

11. Header extensions are intended for limited experimental use. Individual real-time applications normally have their own headers, but these are included in the RTP payload, while the application is identified by its RTP TP as above. Typical application headers are described in Chapter 10. The standard header is 12-bytes long, which together with IP and UDP headers, gives a total header length of 40 bytes. This is much greater than the length of a typical 20-ms burst of low-bit-rate compressed voice, so header compression is essential with this protocol.

A list of payload type assignments is provided in RFC1890 [35].

RTCP

RTCP provides the control functions for RTP through the use of several packet types as follows:

- Sender report gives transmission and reception statistics from participants that are active senders;

- Receiver report gives reception statistics from participants that are not active senders;

- Source description items, including the canonical name (CNAME) (see below);

- Bye message to indicate end of participation;

- Application-specific functions.

The CNAME of a source is required to allow a receiver to combine the RTP sessions for different media, such as audio and video, and to cater for the possible change of SSRC due to conflicting SSRCs or the effect of a reset or restart.

Many RTCP packets are compounds of the separate types above. In particular, reception statistics reports should be sent as frequently as bandwidth availability permits. The RTCP specification provides an algorithm for controlling the frequency of RTCP packets to meet this criterion. Typically the RTCP bandwidth is restricted to 5% of the RTP session bandwidth, with at least a quarter of this reserved for the senders. An interval between RTCP packets is calculated for this, subject to a minimum of 5 seconds, and varied randomly around this figure to prevent unintentional synchronization of reports from multiple participants in the RTP session.

The sender report includes the SSRC of the report originator, timestamps, the sender's packet and octet counts followed by a block of reception reports for each source heard by the sender. These contain the SSRC for each source, the fraction of data packets from that source lost, the cumulative total of such packets lost, the interarrival jitter, and the number of the last sender report and the delay since then.

The receiver reports are similar except for a different payload type and the omission of the sender data. This information is not used directly by RTP, but can be used by senders for an intelligent application to modify their transmissions—for example, by changing RSVP reservations and by receivers and third-party monitors to evaluate performance.

Mixers and Translators

Another function that is offered by RTP for complex network is the provision of mixers and translators.

A mixer is an intermediate system that receives RTP packets from one or more sources, possibly changes the data format, and combines the packets into a new RTP packet, which it then forwards. As the multiple input sources will generally be unsynchronized, the mixer has to make adjustments to synchronize them, and then quote itself as the SSRC. It then puts the original SSRCs in the CSRC list.

A translator is an intermediate system that forwards RTP packets without changing the SSRC or performing any mixing. The main examples are translators from one encoding scheme to another, replicators from multicast to unicast, and some firewall filters. A translator passes through the data streams from distinct sources separately unlike the mixer.

In a 3G network that supports RTP and is connected to other PS networks, it is extremely probable that both these functions will be required. Essentially the mixers and translators connect two or more transport-level network regions. Typically, each such region is characterized by a common network protocol and transport protocol (e.g., IP/UDP), a multicast address or pair of unicast addresses, and a transport level destination port. Mixers and translators must be arranged in such a way that no loops are created. This requires that each of the regions they connect in any given RTP session must either be isolated at the network level or distinct in their protocol/address/port combinations. This is most likely to be applicable to cdma2000 as it is based on Mobility IP and existing standards as much as possible, whereas UMTS does not provide explicit support for these RTP features.

Features that tend to drive the need for these devices are low bandwidth links, differences in end-user functionality, and firewalls. For example, a mixer will allow reduction of the bandwidth for an audio session to that of a single source even when multiple sources are active. A translator might change the video coding from H.323 to the lower bandwidth H.324 for the benefit of mobile participants in a videoconference call or from MPEG-2 to MPEG-4. Such changes would require the translator to change the RTP payload type and provide new sequence numbers and time-stamps for combinations of the simpler video packets. Other probable forms of translation include changing the form of encryption and replacing a multicast address by a set of unicast addresses.

RTP has been fairly widely used to support VoIP as well as with the above-mentioned applications, but it is not proposed for Release 5 of UMTS, although mobiles seem certain to have to interact with it. One of its principal disadvantages is that it does not readily support rate-controlled voice codecs such as AMR, the choice for UMTS. Instead, the preferred applications for UMTS are Session Initiation Protocol (SIP), SCTP, and Media Gateway Control Protocol (Megaco), all of which are outlined in the following sections of this chapter.

7.4.8 SIP

This protocol is defined in RFC2543 [36]. It controls the setup and clear-down of sessions for real-time Internet applications, but not the data transfer nor conference control procedures. It is simpler than the alternative ITU H.245 protocol used for H.323 multimedia (see Section 7.5), and it is preferred to the latter in UMTS (see Section 8.3 and Chapter 10 for examples of this). SIP does not mandate use of IP as the communication protocol for the sessions so may be used for sessions over other protocols, such as ATM or Frame Relay.

There are five basic aspects of SIP operation as follows:

1. *User location:* determination of the end system to be used for communication;

2. *User capabilities:* determination of the media and media parameters to be used;

3. *User availability:* determination of the willingness of the called party to engage in communications;

4. *Call setup:* ringing, establishing of call parameters at both called and calling party;

5. *Call handling:* determination of methods for transfer and termination of calls.

SIP is designed to be able to operate in conjunction with and to support other protocols such as RSVP, RTP, and Real-Time Streaming Protocol (RTSP). It can also operate with various transport level protocols such as TCP, UDP, and SCTP (see Section 7.4.11). When TCP or SCTP is used, then SIP's own reliability features can be dropped.

SIP uses client-server request/response exchanges between application programs in user agents and SIP servers. SIP addressing for callers and callees is based on SIP URLs, which are of the form user@host, or possibly IPv6 addresses. The user part can either be a name or a telephone number (e.g., of the mobile), while the host part is either a domain name or a numerical IP address.

A caller has to locate the nearest SIP server; calling a locally configured proxy SIP server will often do this. SIP requests can then be sent to the SIP server. The main messages associated with a SIP connection are as follows:

- *INVITE:* This invites the callee to join a conference or bipartite call. The content includes a description of the media types and formats likely to be used in a session, doing so in session description language [37]. This request is often sent via a series of proxies and location servers. A response is then sent back by the callee via the proxy to the caller if successful.

- *ACK:* This is sent by the caller to the callee either directly or via the proxy to confirm the successful invitation.
- *BYE:* The user agent of either caller or callee issues this to the server to leave the call.
- *CANCEL:* This is issued to cancel a pending request with the same call-ID and sequence numbers, but it does not affect earlier completed requests.
- *REGISTER:* A client uses this to register an address with a SIP server. One way of doing this is to register on start-up with a local server by sending a REGISTER request to the All SIP Servers multicast address sip.mcast.net or 224.0.1.75, but setting a TTL or scope to limit the propagation of the request to a local administrative area.

In addition to these key messages, there is a range of minor m-types with rough significance indicated by the first numeral in their type identifier as follows:

- *1xx:* informational (e.g., 100 = call trying);
- *2xx:* success, meaning received, understood, and accepted (e.g., 200 OK);
- *3xx:* redirection;
- *4xx:* client error such as failure due to incorrect syntax;
- *5xx:* server error;
- *6xx:* global failure.

An additional type of SIP message is a SIP OPTION that can be used prior to establishment of a connection, for example, to exchange equipment capabilities.

The SIP server can make use of other protocols, such as H.323 for a videoconference, by using their call setup procedures, H.245 for a gateway followed by H.225 in this case, to contact the called party, but this is complicated (see Section 8.1.3). SIP supports both authentication and encryption to provide security. SIP request/responses normally include header fields that carry their names in long-hand; for example, "content-length" precedes the length quoted for the contents of the request/response, but this can potentially lead to the message exceeding the MTU for a low-bandwidth link. To avoid this, SIP offers a compact option, whereby the long-hand name is replaced by an abbreviation (e.g., "l" for content-length or "s" for subject).

The 3GPP, preferencing H.323, has selected SIP for session initiation, and an extended version of SIP will probably meet the specialized needs of UMTS [38].

7.4.9 Effects of Transmission Control Protocol

Transmission Control Protocol (TCP) is a connection-oriented protocol that has been used on top of IP since early in its development [39, 40] to provide reliable transmission in contrast to the connectionless alternative UDP. The reader is assumed to be familiar with TCP, so this section concentrates on the influence that it has on QoS.

TCP performs a range of important functions:

- Splitting of application data into segments to pass to IP that are potentially independent of the read/write size of the application;

- Provision of a checksum for both header and data with dropping of any invalid segments received;

- Setting a timer for acknowledgment of end-to-end receipt of the data segment;

- Acknowledgment of valid data segments received;

- Retransmission of segments if the timer expires before an acknowledgment is received;

- Resequencing of segments that are received out of order;

- Discarding of any duplicate segments received;

- Adjustment of window size by receiver to prevent overflow of its buffers.

Control of these functions is provided through a combination of header fields and the preset TCP protocol operations. The structure of the TCP header is shown in Figure 7.11 in terms of 32-bit words and can be described as follows:

Source port	Destination port
Sequence number	
Acknowledgment number	
Data offset	Reserved	URG, ACK, PSH RST, SYN, FIN	Window size
Checksum	Urgent pointer

Figure 7.11 TCP header.

- The normal header size is 20 bytes but can be extended by as much as another 40 bytes by the optional fields. The source and destination port numbers identify the sending and receiving applications; for example, the familiar Internet File Transfer Protocol (FTP) has port number 23, while RSVP has port number 3328.

- The sequence number is a rolling unsigned 32-bit integer that identifies the first byte of a segment in the data stream being transmitted and wraps around after $2^{32}-1$ bytes, while the acknowledgment number behaves analogously for received data. Selective acknowledgment is not permitted, so if data with sequence numbers 1 to 512 and 1,025 to 1,536 have been received, but not 513 to 1,024, only the first block can be acknowledged, leading to the probable retransmission of 1,025 to 1,536 as well as the missing block.

- The six flags URG, ACK, PSH, RST, SYN, and FYN can be used in combination to indicate the significance of the data. SYN is set by the sender to set up the connection and initialize the sequence numbers, and FYN to end the connection, while RST initiates a reset. ACK indicates that data is being acknowledged, but URG and PSH are less standardized in their usage. The URG flag indicates that the data beginning at the urgent-pointer offset should be processed first, and typically it is used for interrupts and aborts. PSH implies that the data in the segment should be sent up to the receiving application immediately instead of waiting for a full buffer before doing so.

- The 16-bit window size states the maximum number of bytes that the sender is prepared to receive up to a possible 65,535 bytes.

Use of these standard fields alone is not always adequate and has led to the introduction of the optional fields [40]. Each optional field is identified within the optional TCP header extension by a 1-byte kind field, usually followed by a 1-byte length field giving the option header length. The three most important kinds of option for QoS are kind 2 [the maximum segment size (MSS)], kind 3 (the window scale factor), and kind 8 (the timestamp option).

The MSS is only set in the initial SYN segments, with each end stating its preferred MSS. Large values for the MSS up to the MTU of the outgoing interface minus the combined IP and TCP header sizes tend to give the best performance; a default of 536 bytes is assumed if none has been specified.

The window scale factor is important for high-bandwidth routes subject to long propagation delays in order to prevent the window size from being exhausted while waiting for an acknowledgment with consequent interruption to transmission. The window size required for this is given by the formula below, where RTT is the round-trip time:

$$\text{Window size required} = \text{bandwidth} \times \text{RTT} \qquad (7.4)$$

where the bandwidth is that of the worst bottleneck on the route. If this size exceeds 65,535 bytes, then the window size needs to be scaled by a power of two, in order for an adequate size to be specified, with values from 1 to 16 being available by this means. For a 3G mobile user, the most critical case would be trying to achieve 2-Mbps throughput over an international route using one or more satellite hops. Such a route with a single satellite hop in each direction might have an RTT of 600 ms leading to

$$\text{Window size required} = 2{,}000{,}000 \times 0.6/8 = 150{,}000 \text{ bytes}$$

This would require the window size to be scaled by a factor of 4 or more in order to avoid unnecessary interruptions to transmission—but most 3G users will not need this feature.

TCP provides algorithms for calculating the RTT and using it to update the retransmission timer. Use of the optional time-stamps simplifies calculation of the RTT. TCP starts with default values for the retransmission timer and retransmits persistently unacknowledged data after escalating intervals of 1, 3, 6, 12, 24, 48, and 64 seconds. More generally, this back-off procedure uses a progressive doubling of the current value of the retransmission timer instead of the defaults.

TCP also makes use of the Nagle algorithm to prevent cluttering up low-bandwidth links with small packets, and the Slow-Start algorithm to avoid congestion of links or devices that are already potentially heavily loaded at the start of transmission.

The Nagle algorithm [41] is relevant to character mode applications that send a few bytes in each packet wrapped in 40-byte TCP/IP headers, and it states that a TCP connection can only have one outstanding unacknowledged small segment at a time. This leads to buffering of data to form more significant packets with proportionally less overhead. Header compression is an alternative means of alleviating this problem.

The slow-start algorithm states that the sender can only transmit one segment before receiving an acknowledgment at the start of a TCP connection instead of the full window size. After receiving the first acknowledgment, the next transmission can consist of a maximum of two segments, thereafter gradually increasing on receipt of each acknowledgment until the quantity permitted reaches the window size.

Limitations of TCP

TCP is a tremendously successful protocol, but inevitably is not suited to all circumstances. The simplest and most easily rectified issue is the large headers used by TCP/IP. This was recognized early on and largely cured by the Jacobsen header compression [42]. Additional improvements have been made more recently [43–45] that are particularly relevant to mobile phone networks and cover both IPv6 and IPv4. This topic is discussed in the next section.

TCP provides both reliability and data sequencing, but not all applications require both of these. This applies particularly to real-time services where ordering of data can cause unacceptable delays, and limited data dropping may be acceptable. This has led to the development of specialized protocols, such as RTP [34, 35], that run on top of UDP, and also specialized protocols independent of both TCP and UDP, such as SCTP [10]. These are also discussed later in this chapter.

7.4.10   Header Compression

The standard header size for IPv4 is 20 bytes in the absence of source routing, with an additional 8 bytes for UDP or 20 bytes for TCP (excluding options). This can constitute a major overhead where an application generates very small packets. For example, a 20-ms burst of compressed voice at 9.6 Kbps generates only 24 bytes of data, so these additional headers would account for more than half the network traffic and double the network transmission time. Doubling the packet size also doubles the probability of an error during transmission for a given bit error rate, and hence of packet dropping with UDP or retransmission with TCP.

Compression by RFC1144 (Van Jacobsen Compression)

A high proportion of the header is repetitive in nature [e.g., the IP addresses and port numbers (or sockets)] and as such can be abbreviated. RFC1144 defines this procedure for TCP/IP headers for IPv4, but not UDP nor IPv6. The compressor for this standard operates in a simplex manner and generates the following type packets:

- COMPRESSED_TCP;
- UNCOMPRESSED_TCP;
- TYPE_IP.

Packets that do not use TCP are marked as Type_IP and left uncompressed, while some TCP connections may be selected for compression while others are left unchanged. TCP/IP fragments and TCP/IP packets with the SYN, FIN, or RST bits set are also marked as Type_IP and uncompressed since the full header is needed under these circumstances.

The 96-bit header fields consisting of source and destination addresses and ports identify a TCP connection. This combination can be replaced by a connection number, and in the case of RFC1144, this is restricted to a range from 1 to 256 on a compressor and is transmitted and used as an index into an array of header data. Sequence number and window size fields are also shortened by using the difference from those of the previous packet instead of the absolute values, thereby cutting 48 bits to 16 most of the time. The checksum is left untouched. A change mask is also sent to indicate what fields are likely to have changed. The net effect is that 40 bytes are reduced to about 4 to 7 bytes in general.

The low-bandwidth links requiring header compression also tend to be those subject to high error rates, and compression reduces the chances of detecting this condition. The checksum fails to detect quite simple errors, such as two single-bit errors separated by 16 bits, so it is better to rely on the link level and discard frames exhibiting an error. This is not perfect and can lead to incorrect sequence numbers. Error recovery in general depends on sending an uncompressed TCP/IP packet to sort out a confused decompressor. Although it is not ideal for radio links, it is the option quoted for cdma2000 Release A (see Chapter 4 and Section 9.4).

IP Header Compression by RFC2507

IP Header Compression (IPHC) retains the basic compression technique for TCP but improves its error recovery mechanism and extends compression to IPv6 and non-TCP headers, such as UDP; it is one of the options used in UMTS PDCP (see Section 3.3.1).

Traffic is classified as belonging to packet streams characterized by a set of defining fields that include common source/destination address pair and common source/destination port pair. Each such stream is given a context identifier (CID), that is, an 8- or 16-bit number that identifies it. The first packet of the stream is transmitted uncompressed but with the context identifier added. Both the compressor and decompressor store most of these header fields as the context of the packet. Subsequent packets from the compressor transmit only the changes plus the context identifier, and both compressor and decompressor update the context accordingly in addition to their main function. In the case of non-TCP headers, the fields may not change with each packet, so a generation is also defined for these. This is a number from 0 to 63 that is incremented with each packet from the stream. Whenever a field that is expected to remain constant is changed, a full uncompressed header is transmitted.

Four types of packet are transmitted for each of IPv4 and IPv6 as follows:

- FULL_HEADER: This has an uncompressed header with the addition of a CID, and also a generation for non-TCP packets.

- COMPRESSED_NON-TCP: This has a compressed header that includes the CID, generation, and the randomly changing header fields.

- COMPRESSED_TCP: This has a compressed TCP header that includes the CID, an octet flag field indicating which fields have changed, and the differences of these fields from their last value.

- COMPRESSED_TCP_NODELTA: This is a packet with compressed TCP header except that the full values of the changed fields are sent instead of the changes. It is only sent in response to a header request from the decompressor.

In addition to these, a decompressor can also send the compressor a CONTEXT_STATE packet that contains a list of CIDs for TCP streams for which synchronization has been lost. It is only sent over a single link and contains no IP header. As with RFC1144, the compressed TCP headers are usually 4 to 7 bytes long. A reliable link is required, both for IPv6 in general and for IPv4 with header compression, for this method. A link level protocol can detect damaged frames but cannot reliably identify the packet stream as it may be the CID that has an error; so it is best to transmit a strong checksum in the compressed header for IPv4.

Error recovery can use the same techniques as RFC1144, but also has improvements. Thus, if a TCP segment is lost, the decompressor will detect that checksums for subsequent packets for that CID are wrong, and so fail to acknowledge them with resultant retransmission of a full TCP window. This could happen quite frequently on a radio link, so a better mechanism is needed.

One such method is the so-called Twice algorithm where the decompressor guesses that the deltas for the missing segment were the same as those for the current segment and tries adding twice the deltas to see if it gets a checksum match. If so, it reconstructs the missing packet and carries on. If not, it tries once more before giving up. This technique is found to work well more than 50% of the time [43]. The Twice algorithm is less effective for acknowledgments than for source data packets. The second recovery method for TCP is for the decompressor to send a CONTEXT_STATE, ideally for more than one CID to avoid cluttering the link with numerous requests.

In the case of non-TCP packets, a compressed header with an out-of-order generation number can either be discarded immediately or held until an uncompressed packet with the correct generation number has been received, subject to a time-out to prevent wrap-around issues for the 6-bit generation number.

IPHC does not directly handle QoS issues, but the IPv4 TOS and IPv6 traffic classes can be used as defining fields for the context identifier so that streams of packets with distinct QoS needs within a single application can be treated separately, for example, in relation to recovery technique.

RFC2507 does not define compression of RTP headers, but an extension, called compressed RTP (CRPT), is defined in RFC2508 to do this. This reduces the 40-byte IP/RTP/UDP header most of the time to either 4 bytes if UDP checksums are sent or 2 bytes if not. This very high degree of compression is possible because many of the fields change by a constant amount or not at all in successive packets, allowing all to be eliminated by use of the differences of differences. From time to time a major field does change, and under these circumstances, a much fuller header has to be sent. Unfortunately, this algorithm does not work sufficiently well at the high error rates encountered on mobile phone networks. Even with the Twice algorithm, it is still possible to get gaps in transmission of the order of 100 ms, which is not good enough for real-time multimedia, in addition to which large headers have to be sent after errors. Accordingly, another more robust compression algorithm has been produced.

Robust Header Compression

Robust header compression (ROHC) is a much more complex algorithm designed to meet the needs of real-time services over error-prone networks [45] and to be used in the UMTS PDCP function (see Section 3.3.1). It supports four header profiles, RTP/UDP/IP, UDP/IP, ESP/IP (encapsulated security payload), and uncompressed, but not TCP or SCTP. It supports both IPv4 and IPv6, including its optional header extensions.

For RTP, the basis of ROHC is the recognition that many of the variable fields can be predicted on the basis of the RTP sequence number. Consequently, ROHC first works out the functional relationship of these fields with the sequence number and transmits the latter.

In addition to the robust error handling, ROHC also addresses the issue of supporting the use of both multiple channels and shared channels in a 3G network for a single call by means of CID. The compressor and decompressor each maintain the context of the compressed flow—based on the state of the fields sent—in order to perform correctly, and the CID value space is specific to an individual channel, so that identical CID values for different channels refer to distinct contexts.

ROHC makes use of three distinct operational states:

1. *Initialize and refresh (IR):* In this state, which occurs at the start or after a failure, the compressor sends complete headers to the decompressor to initialize the static fields.

2. *First Order (FO):* Once the compressor has had sufficient ACKs from the decompressor, it moves to the FO state for communication of packet irregularities. In this state, the information is at least partially compressed.

3. *Second Order (SO):* This is the optimal state, which is reached once the compressor has reason to believe that the decompressor has full knowledge of the relationship between the sequence number and other fields.

Analogously, the decompressor has three states: no context, static context, and full context.

ROHC also has three modes of operation, each of which uses all three states, as follows:

1. *Unidirectional (U-mode):* Data is sent in one direction only—that from compressor to decompressor. This is the least efficient mode, and all sessions start this way, with transition to one of the other modes possible after a suitable feedback from the decompressor.

2. *Bidirectional optimistic mode (O-mode):* The main difference of this to U-mode is the limited use of a feedback channel. The aim is to maximize compression but at the possible risk of context invalidation at high error rates.

3. *Bidirectional reliable mode (R-mode):* This bidirectional mode makes extensive use of the feedback channel to provide maximum robustness.

There are a number of techniques used in carrying out the compression. In addition to recognizing fixed fields, one of the key features for real-time traffic is the use of stride factors for the regular increment of RTP time-stamp for audio and video streams. Variation in the received times for such packets allows the decompressor to estimate the jitter. There is a range of packet types sent from compressor to decompressor that depends on each of the state, mode, and traffic types.

The ability of ROHC to work over unreliable links depends on a comprehensive set of feedback mechanisms from decompressor to compressor. There are three basic types of feedback, ACK, NACK, and STATIC-NACK, but each of these can contain additional feedback in option fields. The NACK indicates a synchronization error in the dynamic context of the decompressor, while a STATIC-NACK shows that the static context is invalid or has not been established. The optional fields include a CRC field for the feedback, lack of decompressor resources, sequence number information or errors, clock sensitivity of the decompressor, maximum jitter in a period, and the largest number of packets seen to be lost in a sequence.

7.4.11 Stream Control Transmission Protocol

SCTP is a reliable protocol defined in RFC2960 that runs on top of connectionless protocols such as IP and that is more suited to real-time streamed traffic than TCP. Its primary function is to convey PSTN signaling across IP networks—but it can also be used for data—and is specified in Release 5 of the UMTS specifications for interface to IP multimedia core networks. The main characteristics of SCTP are as follows:

- Acknowledged error-free nonduplicated transfer of user data;
- Data fragmentation to conform to discovered path MTU size;
- Sequenced delivery of user messages within multiple streams, with an option for order-of-arrival delivery of individual user messages;
- Optional bundling of multiple user messages into a single SCTP packet;
- Network-level fault tolerance through supporting of multihoming at either or both ends of an association;
- More resistance to SYN style attacks than TCP.

The most significant difference from TCP is the support for multihoming, with the ability to handle congestion separately on the different paths between the end points. It also allows faster retransmit and recovery from errors than TCP. The following paragraphs cover the aspects of SCTP most closely associated with congestion and flow control only.

An SCTP packet consists of a 12-byte common header (as in Figure 7.12), followed by one or more data chunks. A data chunk may be either control data or user data, and SCTP has a bundling option to enable messages from more than one user to be combined in a single SCTP packet as separate chunks. The common header fields are as follows:

- *Source port number* (16 bits): The SCTP port number of the sender;
- *Destination port number* (16 bits): SCTP port number of receiver —these two SCTP port numbers can be used together with the

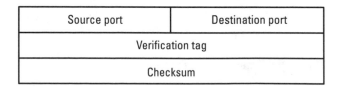

Figure 7.12 SCTP common header.

respective IP addresses to define the association to which the packet belongs;

- *Verification tag* (32 bits): The receiver to validate the sender uses this;
- *Checksum* (32 bits): It covers the SCTP packet.

The chunks, each of which has its own 4-byte chunk header, follow the common header. The chunk header fields are as follows:

- *Chunk type* (8 bits): Values range from 0 to 254. Notable examples include 0 for payload data, 1 for initiation, 2 for Init Ack, 3 for selective acknowledgment (SACK), and 7 for shutdown. The two highest-order bits say what to do if the receiver does not recognize the chunk type; 00 means discard it;
- *Chunk flags* (8 bits): The use of these depends on the chunk type. Usually they are set to zero on transmit, and ignored by the receiver;
- *Chunk length* (16 bits): This is the combined length of the chunk header and chunk value field in bytes, excluding any padding;

The chunk value field itself must be a multiple of 4 bytes long, and it is padded with up to 3 bytes if necessary to ensure this. Chunks for control purposes have a standard structure consisting of the following:

- *Parameter type* (16 bits): Identifier of type of parameter with range 0 to 65,534;
- *Parameter length* (16 bits): Length in bytes of the parameter including type and length;
- *Parameter value* (variable length): The information to be transferred by the parameter.

The data payload has its own headers that govern flow and congestion control. The mandatory format is shown in Figure 7.13.

The significance of these fields is as follows:

- *Type* (8 bits): Zero for the data payload;
- *Reserved* (5 bits): Currently set to zero and ignored;

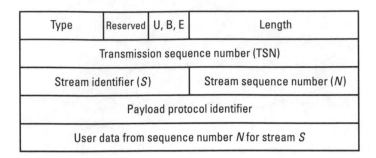

Figure 7.13 SCTP data chunk.

- *U-bit* (1 bit): Set to 1 to indicate data chunk is unordered and any stream sequence number is to be ignored;

- *B-bit* (1 bit): Set to 1 to indicate the first fragment of a user message;

- *E-bit* (1 bit): Set to 1 to indicate last fragment of a user message (an unfragmented message has both flags set);

- *Length* (16 bits): The length of the data chunk in bytes including these header fields but excluding any padding;

- *Transmission sequence number (TSN)* (32 bits): Required to reconstruct a data message from its data chunks and has values from 0 to $2^{32}-1$, with automatic wrap-around. It is independent of the stream sequence numbers;

- *Stream identifier* (16 bits): Identifies stream to which the following user data belongs. A single association may have several separate message streams (e.g., audio and video) with further subdivisions;

- *Stream sequence number* (16 bits): The stream sequence number for the following data from that stream. The same for all data belonging to the same fragment where SCTP fragments data packets for transmission to match the path MTU;

- *Payload protocol identifier* (32 bits): Not used by SCTP itself, but passed to upper layers and peer;

- *User data* (variable length): The actual user data.

Data Transmission in SCTP

SCTP will both bundle small messages and fragment large packets as required for efficient transmission. Flow control is based on window size information, selective acknowledgments, and timers for retransmission. Window sizes are initially set via the INIT and INIT ACK messages, with a minimum of 1,500 bytes for the receive window. Acknowledgments are sent in SACK chunks or SHUTDOWN if appropriate.

The initial receiver window, called the advertised receiver window credit (a-rwnd) in the INIT is a 32-bit number of bytes of buffer space that the sender has reserved for the window, with a similar parameter returned in the INIT ACK.

The amount of data that can be sent before receiving an acknowledgment is given by the congestion window (cwnd). A sender cannot send any further data other than cwnd if rwnd is zero. The difference in roles of these two parameters is that rwnd describes the buffer space available for the entire association (including multiple addresses for a multihomed receiver), whereas cwnd controls congestion on a per-address basis. The initial value for cwnd is equal to or less than twice the MTU and is automatically reduced to not more than the MTU if retransmission is required.

The SACK is much more versatile than a TCP acknowledgment. It contains not only a cumulative TSN field to acknowledge all TSNs up to and including that value, but also gap acknowledgment blocks. This is done by having a 16-bit field to indicate the number of such blocks, then for each such block there is a 16-bit block start offset relative to the cumulative TSN and a block end offset relative to the cumulative TSN. This enables the sender to retransmit unacknowledged blocks selectively, although the SACK also has fields for acknowledging duplicate TSNs individually. Unlike TCP, SCTP can pass sequences of blocks up to the higher layer protocol without waiting for all the gaps to be filled; this is particularly helpful for applications that do not require ordered data. The SACK also carries an updated rwnd.

SCTP uses delayed acknowledgments to eliminate unnecessary traffic. The delay is configurable, typically to 200 ms, subject to a maximum of 500 ms, and the cumulative TSN used to acknowledge all chunks received up to the time-out. If, however, a gap is found, then the receiver sends a SACK for every chunk received. The sender keeps a count of the number of times each gap is reported in a SACK and marks a chunk for retransmission when this count reaches 4, while also starting the fast retransmit procedure.

This entails setting a slow start threshold (ssthresh) and the cwnd in accordance with specific rules, but crudely to half the normal cwnd. As many missing data chunks as possible are fitted into an SCTP packet, subject to the MTU, starting with the lowest TSN, and this is transmitted. Once a SACK is received for this, the cwnd can be increased in line with specified slow start procedures.

Path failures or destination idleness are detected on the basis of an error count exceeding a specified value. This count is incremented every time a retransmit time-out is exceeded and when a keep-alive heartbeat is not returned within another time-out.

7.4.12 Megaco

This protocol [46] has been designed to support the need for the media gateways with distributed subcomponents that are often required in complex networks. It is specified in RFC3015 and aligned with ITU-T specification H.248, which itself supplements the earlier H.245 gateway component of the H.323 videoconferencing standard. The provision of this as an open standard means that Telcos and others can now purchase their media gateways and gateway controllers from different vendors with potentially lower costs as a result. It also allows growth to take place by the addition of extra media gateways of a common type running under the same controller, instead of replacing low-capacity gateways by high-capacity specimens.

A 3G mobile network will have requirements to interface to external services over a variety of media and using formats that require conversion to enable effective communication. The Megaco Protocol provides a framework for the operation of media gateways (MG) and specifies how they interact with a media gateway controller (MGC) for connection control. Section 8.1 mentions examples of this in UMTS.

The basic concepts used to define connection control are terminations and contexts.

A termination acts as the source or sink for one or more media streams or control streams, while a context is an association of a number of terminations. The context defines who sees or hears whom and also covers any mixing or switching parameters that are required between different terminations. The number of terminations per context is a characteristic of an individual media gateway. An MG that handles access and conversion for point-to-point links may be restricted to two terminations, whereas an MG for multipoint processes will normally have to support at least three.

A termination can either represent a physical entity, such as a trunk interface port or channel on that trunk, or an information flow, such as RTP. A termination has two main features as follows:

- *Termination ID:* This is an identifier issued by the media gateway according to its own scheme when the termination is created and may be structured, for example, to indicate channels within a trunk.

- *Properties and descriptors:* The characteristics of a termination are given as properties that have an ID and a description. These properties and their descriptions are listed in Table 7.7.

Descriptor Formats

Megaco provides two styles of expression for the descriptors: textual or binary-encoded. The textual format uses abbreviations of the field names based on System Description Protocol (SDP) rules, while the binary option uses concise tag values to express the Property ID for local/remote descriptors accompanied by binary tag values in defined field sizes and format. The Property ID tag values are divided into specific groups according to the type of descriptor, as indicated in Table 7.8.

Table 7.7
Termination Properties

Descriptor	Description
Modem	The type of modem, if any (e.g., V.90)
Mux	The type of multiplexing, if any (e.g., H.221/223/225, AAL2)
Media	Media stream descriptions comprising termination state descriptor (TSD) plus stream ID
TSD	Nonstream-specific aspects of termination (e.g., in service)
Stream ID	Used to link streams in a context, subdivided into local control, local, and remote
Local	Properties that specify the media stream received by MG
Remote	Properties that specify the media stream sent by the MG
Local control	Properties of interest between MG and MGC
Events	Events to be detected by MG and required action
Event buffer	As previous, but when event buffering in use
Signals	Signals and actions (e.g., busy tone) to be applied to termination
Audit	Information solicited in audit commands
Packages	List of packages realized by termination
Digit map	Patterns to enable sets of events to be reported as a group event
Service change	Reasons for a service change
Observed events	Report of events observed
Statistics	Report of statistics kept on a termination

Table 7.8
Property ID Tag Value Ranges

Nature of Property	Tag Range
General media attributes	1000–10FF
Mux properties	2000–20FF
General bearer properties	3000–30FF
General ATM properties	4000–40FF
Frame-relay properties	5000–50FF
IP properties	6000–60FF
ATM AAL2 properties	7000–70FF
ATM AAL1 properties	8000–80FF
Bearer properties (mainly Q.931)	9000–90FF
ATM AAL5	A000–A0FF
SDP equivalents	B000–B0FF
H.245 attributes	C000–C0FF

The parameter values associated with these tags define the characteristics of the media streams sent or received by the media gateway. Some of the types of specific importance to 3G networks are described in the next chapter, notably general media and AAL2 and AAL5 attributes.

Megaco Commands

The MGC controls its media gateways by means of the following set of commands:

- *ADD:* This adds a new termination to an existing context.
- *MODIFY:* This modifies the properties, events, or signals of a termination.
- *SUBTRACT:* This disconnects a termination from a context, and returns statistics on its participation.
- *MOVE:* This moves a termination from one context to another.
- *AUDIT VALUE:* This returns the current state of events, signals, properties, and statistics of a termination.
- *AUDIT CAPBILITIES:* Returns all possible values for termination properties, events, and signals allowed by the media gateway.

- *NOTIFY:* Allows the media gateway to inform MGC about events.

- *SERVICE CHANGE:* This allows an MG to announce its availability or restart to the MGC, and to notify the MGC of a change in service status of a termination.

Operation

The Megaco standard does not specify the network level communications standard between the MGC and MGs, and initial implementations are still proprietary in this regard, although Q.BICC is likely to develop into the open standard for this. Currently, either TCP or UDP/ALF is recommended for the transport layer if IP is used at level 3.

The standard recommends starting MGs at different times to prevent overload of the MGC by all doing so at once. This is achieved by using a random delay after power-on that lies in a range from zero up to a maximum waiting delay (MWD). Suggested vales for the MWD range from about 60 ms for an MG on a T3 trunk up to 600 ms for an MG on an analog modem link, reflecting the relative frequency of operations required.

Media gateways are potentially liable to denial-of-service attacks through overload caused by source-spoofing. The use of IPSec is specified to prevent this with IPv6 with the option of the earlier interim authorized header scheme for IPv4.

7.4.13 MPLS and IP over ATM

QoS based on RSVP and Intserv in IP or virtual circuits in ATM requires a great deal of processing and configuration, as well as delays in rerouting if a major node fails. As a result, it is very unscalable.

Multiprotocol Label Switching (MPLS) [47] provides a much more scalable approach, and its probable use in UMTS and cdma2000 is mentioned in Chapter 9. The complexity of an ATM or an Intserv network is proportional to the square of the number of nodes, whereas that of an MPLS network only rises linearly and is more suited to the trend toward any-any routing in place of the traditional hub-spoke architectures. The key feature is the observation that a router effectively divides incoming packets into forwarding equivalence classes (FECs) and then forwards them to the next hop on that basis only. MPLS formalizes this procedure by working out these equivalence classes on entry to the network and assigning them a label on that basis that is then carried through the network to use for routing. The label simply provides an index into a table of next hops. Assignment of labels can be made on both the contents of the packet header and other information, such as the entry port on the router that could differentiate between distinct source networks subject to separate tariffing. MPLS allows routing to be done either on a hop-by-hop basis per FEC by each node independently as in most IP networks or through explicit routing. In the latter case, the ingress (or occasionally the egress) label switch router (LSR) specifies several (or all) of the LSRs on the label switch path (LSP).

The labels have a local significance within the network based on binding agreement between an upstream and downstream LSR on the relation between a given label and an FEC. A packet may carry multiple labels in a label switch stack related to different bindings, and processing of a packet is based on the label at the top of the stack. It is the downstream member of a pair that decides to bind a particular label to a particular FEC. One of several different possible Label Distribution Protocols (LDPs) [48] is used to distribute the bindings between LSRs—for example, MPLS-BGP, MPLS-RSVP-TUNNELS, or MPLS-CR-LDP, of which the first two examples make use of extensions to BGP and RSVP, respectively, while the third is a new development.

The label stack has a last-in first-out (LIFO) structure. If a nonempty stack is of depth m, then the top label has level m and the bottom has level 1. The main role of the label stack is to support nested sets of LSP tunnels. A nominal LSP R_1, R_2, R_3, R_4 characterized by the four LSRs may contain indirect tunnels via other LSRs; for example, the first hop from R_1 to R_2 may be direct and use a level 1 label, but from R_2 a packet may go through another sequence of j LSRs, R_{21}, ..., R_{2j} in order to reach R_3. If so, then R_2 puts a level 2 label onto the stack for use by the intermediate LSRs, the last of which removes the level 2 label, so that level 1 is again used on the final hop.

This use of labels bears some resemblance to the operation of an ATM switch where the entry port and VCI/VPI are used as an index into a switching table. It is thus only a small change to use an ATM switch as a label switching router for MPLS, but with much improved scalability due to use of equivalence classes instead of PTP VCs.

The basic specification of MPLS [47] does not address QoS factors, but recent extensions enable it to support TOS in IPv4, RSVP, and more importantly, DiffServ. Most of the traffic on an IP network is likely to be of traditional types that are tolerant of both delays and errors, with only a small amount of intolerant real-time applications. As a result, use of DS classes EF and AF12 for the real-time applications (with the rest on a best-effort level such as AF11), accompanied by traffic engineering [49] and policing, should give adequate performance across the network. Recent enhancements

(MPLS-RSVP-TUNNELS and MPLS-CR-LDP) also allow for QoS-dependent routing, using explicit routing. In MPLS-RSVP-TUNNELS [26] the existing RSVP mechanisms (see Section 7.4.4) are extended to support label distribution and explicit routing. In this case the RSVP FLOWSPECS are used to define the labels, and this use of labels then avoids the need to apply an RSVP FILTERSPEC at each node. Five new RSVP objects are defined in addition to those in Table 7.3: LABEL_REQUEST (RSVP object class 19), LABEL (16), EXPLICIT_ROUTE (20), RECORD_ROUTE (21), and SESSION_ATTRIBUTE (207) associated with PATH and/or RESV messages. Furthermore, new c-types are added to the existing SESSION, SENDER_TEMPLATE, and FILTER_SPEC object classes to apply to LSP tunnels using IPv4 or IPv6 addresses. The EXPLICIT_ROUTE object is the feature most directly relevant to control of QoS. It allows predetermined routes to be created as a result of traffic engineering and used instead of normal IP routing for unicast traffic. The route is specified in terms of subobjects that can represent specific nodes with IPv4 or IPv6 addresses, groups of nodes (to allow more flexibility of routing), or autonomous systems. There is also a hello message to allow rapid detection of node failures, while the RECORD_ROUTE permits discovery of any routing loops.

MPLS-CR-LDP [50] adds resource reservation procedures to the existing MPLS explicit routing and label distribution techniques. It uses requirements for peak and committed data rates and burst sizes in selecting the route. The route selected will be the shortest path that meets these constraints, rather than the shortest path in general.

7.4.14 Session Description Protocol

Many of the preceding RFCs, notably SIP, and some of the 3G literature make use of SDP [37] to describe their features. This protocol is outlined here, and examples of its use included in Chapter 10. SDP is intended for use in multimedia session initiation and control, particularly for the advertisement of conference sessions and their setup, and is widely used for this. Information is provided in a textual format mainly using ISO10646 characters in UTF-8 encoding [51] to provide a compact presentation. Information is provided for any of the range of fields listed below according to their single letter identifier (the items with an asterisk are optional):

- v = protocol version;
- o = owner/creator and session identifier;

- s = session name;
- $i*$ = session information;
- u = URI of description;
- $e*$ = e-mail address;
- $p*$ = phone number;
- $c*$ = connection information—not required if included in all media;
- $b*$ = bandwidth information;
- One or more time descriptions (see below);
- $z*$ = time zone adjustments;
- $k*$ = encryption key;
- $a*$ = zero or more session attribute lines;
- Zero or more media descriptions (see below).

Time description:

- t = time the session is active;
- $r*$ = zero or more repeat times.

Media description:

- m = media name and transport address;
- $i*$ = media title;
- $c*$ = connection information—optional if included at session-level;
- $b*$ = bandwidth information;
- $k*$ = encryption key;
- $a*$ = zero or more media attribute lines.

For the multimedia sessions over 3G networks for which QoS is most essential, one of the key lines is that for the media name and transport address. Frequently, a session will entail several different data streams, such as one each for video, audio, and control; some hypothetical examples are given below. Most of these sessions are carried in RTP/UDP/IP, so typical lines are

$$m = \text{video } n \text{ RTP/AVP } x$$

where video indicates a video application, n denotes the numeric UDP port number for the RTP packets, RTP/AVP indicates RTP audio or visual, and x

indicates the type of audio-video stream as the RTP payload type (e.g., 31 for H.261);

$$m = \text{audio } n \text{ RTP/AVP } x$$

for an audio stream (e.g., with a value 3 for x indicating GSM voice);

$$m = \text{control } n \text{ H.323 } mc$$

indicates a control stream for an H.323 multipoint control unit using UDP port n. Where RTP is involved with a quoted UDP port number of n, there is an associated RTCP stream with UDP port number $n + 1$. Where the optional bandwidth attribute b is used, the value is given in kilobits per second.

UMTS requires the use of some of these features, while leaving others optional [52]. Mandatory fields include protocol version, owner and session name for the session, media name, transport address, connection information, and bandwidth for the media description plus a set of attributes. The attributes that are required are control, RTP map, range, and format type. A single media feature, such as audio, may use several different RTP payload types to a single UDP port for distinct traffic streams representing different codecs, in which case the m statement lists all the RTP payload types, and a statements are required for each RTP payload type; for example,

$$m = \text{audio } n \text{ RTP/AVP 96 97}$$

$$a = \text{rtpmap 96:L8/8000}$$

$$a = \text{rtpmap 97:L16/8000}$$

for 8-bit and 16-bit voice codecs with 8,000-Hz sampling rates. The format type typically distinguishes between different profiles and levels that can be used within a single multimedia application [e.g., MPEG-4 (see Section 10.2.6) or the modes used in AMR voice (see Section 10.1.1)]. The control and range attributes are needed when SDP is used to describe a RTSP session [53], where range describes the total time for a stored session. Numerous optional attributes exist (e.g., recvonly for receive only, and sendrecv for send and receive).

7.4.15　Mobility IP

Cdma2000 makes extensive use of Mobility IP (see [54] and Section 8.2.3) rather than separate cellular radio standards for both control of mobility and interaction with fixed IP networks. This standard was created to allow transparent routing of datagrams to and from mobile IPv4 nodes, regardless of their location. The mobile station always uses its home address, but also uses a care-of address that is registered with a home agent to enable messages to reach it. The home agent is a router on the mobile's home network that encapsulates datagrams for the mobile in a header that uses the care-of address. There are two types of care-of addresses: a foreign agent and a colocated address. The foreign agent is a router that the mobile has registered with on its temporary network; the colocated address is an IP address temporarily acquired by the mobile on one of its interfaces, either dynamically by DHCP or from a permanent pool.

Mobility agents make use of agent advertisement to advise mobiles of their presence and to enable the latter to determine their current point of attachment. These advertisements are extensions to the normal ICMP router advertisements. Agents may also be discovered by link level protocols.

Mobiles send agent solicitations analogous to ICMP router solicitations if they do not receive any agent advertisements. A mobile may send up to three solicitations at 1-second intervals initially, but thereafter the rate must be reduced.

The protocol contains two control messages (registration request and registration reply) that are sent on UDP port 434. It also makes use of ICMP router discovery. Message extensions are further defined in Mobility IP for each of these types of message to enable mobile-home authentication, mobile-foreign authentication, and foreign-home authentication.

Once a mobile has registered with a foreign agent, it selects its default router from the list advertised by that agent.

Mobility IP does not specify any QoS mechanism, but it is normally used with differentiated service. RSVP is not suited to the triangular routing wherein a mobile sends to the destination via its foreign agent, but receives by a path via both home and foreign agents.

7.5　H.323

ITU-T recommendation H.323 [55] was the first and most widely used standard for control of multimedia traffic streams over PS networks (especially LANs) that do not guarantee QoS. It has undergone numerous modifications, and the fifth version was due in 2002. It is an umbrella standard that references several other standards for specific functions, notably H.225 registration, multiplexing, and signaling, and H.245 call control. Each version of H.323 also specifies a specific version of each of these other standards; for example, H.323v4 specifies version 4 of H.225.0 and version 7 of H.245.

H.323 provides recommended standards for four classes of equipment type: terminals, gateways, gatekeepers, and multipoint control units (MCUs). Only the terminals are always essential; the need for the others depends on the complexity and function of the network. Terminals are the user end points and must support specific voice codecs for the ITU recommendations G.711, G.722, G.723, G.728, and G.729, but the main mobile phone compression standards (see Section 10.1) are not included. Support for video is not mandatory, but where it is supported, the terminal should have a codec for the H.261 recommendation and optionally also for H.263 (see Chapter 10 for these codecs). There is no support for the MPEG multimedia codecs (see Chapter 10) except for a limited degree of commonality between MPEG-4 and H.263. Support for data conferencing by ITU T.120 is also recommended in later versions of H.323. The terminals must also support H.225 and H.245. UMTS does not recommend its mobiles to be H.323 terminals and specifies SIP [36] for multimedia session control instead; however, both UMTS and cdma2000 devices are likely to need to communicate with H.323 terminals. Section 8.1.3 discusses the main issues involved in this.

Gateways are an optional feature that may be required to enable communication with systems using other standards. The standards envisaged by the ITU are those of its own recommendations for services on other types of network, such as PSTN, ISDN, and B-ISDN. Thus, a gateway should enable translation between H.225 and H.223/H.221 and between H.245 and H.242 in particular. In addition, transcoders may be provided for interworking with terminals using other codecs. H.323v4 is compatible with the use of ITU H.248 [46] for the control of multiple media gateways. Interworking between UMTS and H.323 requires a gateway that can translate between SIP and H.225/245. An H.323 terminal uses H.245 operations to communicate with a gateway.

Gatekeepers are optional devices for the control of individual zones and are a key feature of most H.323-enabled networks. They provide call control for terminals, gateways, and MCUs registered within their own zone and act as virtual switches for calls, with call routing as an optional feature. Each zone (frequently one or more LANs) has a single gatekeeper, so gatekeeper functionality can be disabled in multifunctional H.323 devices. In H.323v2 and more specifically in H.323v4, provision is made for alternate gatekeepers to provide resilience. The functions provided by the gatekeeper are the registration, admission, and status (RAS) range of address translation, admissions control, bandwidth management, and zone control. H.225 defines the RAS operations and also contains a signaling component that is based on the ISDN Q.931 recommendation. RAS operates between a terminal and the gatekeeper, while the gatekeeper uses the Q.931 element to set up connections from terminal to terminal.

Address translation includes both translations between terminal aliases and IP addresses and those between E.164 PSTN/ISDN addresses and IP addresses. The number of types of addresses varies according to H.323 version, increasing from only two in version 1 to five by version 4 (E164ID, H323ID, E-MailID, URL ID, Transport ID). Admission control governs access to the LANs of H.323 terminals by means of ARQ, reject, and confirm messages based on administrative and bandwidth availability considerations. Bandwidth control is based on bandwidth request, reject, and confirm messages, and is used to prevent the H.323 terminals from using up too much bandwidth rather than to try to provide QoS for them.

The fourth type of H.323 device is the MCU, which is required whenever three or more H.323 terminals participate in a conference call. The MCU consists of a single multipoint controller (MC) to which multipoint processors (MP) may optionally be added. The MC uses H.245 messages to determine the capabilities of the participating terminals and negotiate the features to be used, in addition to deciding what can be multicast. The MC does not process the media streams, and if required this must be performed by the MP. These actions of the MP comprise mixing, switching, and processing of the contents (e.g., codec-to-codec conversion) for the media streams. Conference control can be performed centrally by the MCU or by decentralized multicasts or by hybrid control where some streams (e.g., audio) may be handled centrally and others, like video, may be multicast.

Media streams, signaling, and call control messages are all carried within IP normally. Those functions for which accuracy is essential (i.e., control, signaling, and T.120 data) are carried within TCP, while those for which urgency is more important than reliability (e.g., audio and video) are carried in RTP/UDP [34]. A single terminal-to-terminal call in H.323 requires two separate TCP sessions. The first, for the initial H.225/Q931 call, goes to a well-known port number. The response includes a dynamic port number for the subsequent H.245 conference control and capability exchange. Once this channel has been opened, the first TCP session can be closed.

Although H.323 has no direct support for QoS, some requirements, such as sequencing and jitter control, are provided through the use of RTP time-stamps by the terminals. Audio and video channels under H.225.0 use separate RTP and RTCP sessions and prior to H.323v4 could only be synchronized with each other by the RTP time-stamps. This sometimes proved difficult, so H.323v4 allows for multiplexing of the two. There is a single H.245 logical channel between each pair of terminals for control of the media stream sessions. They are particularly important for video and cover such functions as freeze frame, rewind, frame synchronization, and fast forward. H.245 is used for capability exchange using descriptor values that provide an index into preset tables of capabilities for codecs and bit rates.

From version 2 onwards, H.323 also makes use of H.235 for security features, (principally authentication) and H.450 for supplementary services (e.g., call hold, redirect). The number of supplementary services has increased in each version. Specific recommendations also exist for interworking with ATM networks [15] using AAL5 cells for video and either AAL5 or AAL2 for audio. Some early H.323 devices also have support for RSVP by proprietary means, and H.323v4 introduces some features to allow for RSVP.

References

[1]   ETSI TS 125 101, "WCDMA-FDD UE Radio Transmission and Reception," V.4.2.0 (2001), http://www.etsi.org.

[2]   ETSI TS 25.410, "UTRAN I_u Interface: General Aspects and Principles," V.4.2.0 (2001), http://www.etsi.org.

[3]   ETSI TS 25.412, "UTRAN I_u Interface: Signaling Transport," V.4.0.0 (2001), http://www.etsi.org.

[4]   ETSI TS 25.414, "UTRAN I_u Interface Data Transport and Signaling," V.4.1.0 (2001), http://www.etsi.org.

[5]   Russell T., *Signaling System # 7*, 2nd ed., New York: McGraw-Hill, 1998.

[6]   Knight, D., and B. Law, "Introduction to Signaling," in *Broadband Signaling Explained*, pp. 23–42, D. Knight (ed.), New York: John Wiley & Sons, 2000.

[7]   ITU-T Recommendation Q.711 "Functional Description of Signaling Connection Control Part," 1996.

[8]   ITU-T Recommendation Q.2150.1 "B-ISDN ATM Signaling Transport Converter for MTP-3b," 1996.

[9]   ITU-T Recommendation Q.2210 "MTP-3b Functions and Messages Using Q.2140," 1996.

[10] IETF RFC2960, "Stream Control Transmission Protocol (SCTP)," 2000.

[11] ITU-T Recommendation Q.2140 "B-ISDN ATM Adaptation Layer Service Specific Network Network Interface (SSCF-NNI)," 1995.

[12] ITU-T Recommendation Q.2110 "B-ISDN ATM Adaptation Layer Service Specific Connection Oriented Protocol," 1996.

[13] ITU-T Recommendation I.363.2 "AAL2 Adaptation Layer" 1996.

[14] ITU-T Recommendation Q.2630.2 "AAL2 Signaling Protocol (Capability Set 2)," 2000.

[15] http://www.atmforum.org.

[16] ITU-T Recommendation I.371 "ATM Service Categories," 1996.

[17] IETF RFC2131 "Dynamic Host Control Protocol (DHCP)," 1997.

[18] IETF RFC2460, "IP Version 6 Draft Standard," 1998.

[19] IETF RFC2462, "IP Stateless Address Configuration," 1998.

[20] IETF RFC2210, "Integrated Services," 1997.

[21] IETF RFC2211, "Integrated Services Controlled Load," 1997.

[22] IETF RFC2212, "Integrated Services Guaranteed Quality of Service," 1997.

[23] IETF RFC2205, "Resource Reservation Protocol (RSVP)," 1997.

[24] IETF RFC2750, "RSVP Extensions for COPS," 2000.

[25] IETF RFC2207, "RSVP Extensions for IPSec," 1997.

[26] IETF RFC3209, "RSVP-TE: Extensions to RSVP for LSP Tunnels," 2001.

[27] IETF RFC2474, "Definition of Differentiated Services for IPv4 and v6," 1998.

[28] IETF RFC1812, "IPv4 Routing," 1996.

[29] IETF RFC2597, "DiffServ Assured Forwarding PHB," 1999.

[30] IETF RFC2598, "DiffServ Expedited Forwarding PHB," 1999.

[31] IETF RFC2998, "IntServ over DiffServ," 2000.

[32] IETF RFC2748, "COPS Common Open Policy Server," 2000.

[33] IETF RFC2753, "Framework for Policy Based Admission Control," 2000.

[34] IETF RFC1889, "Real-Time Transport Protocol (RTP)," 1996.

[35]   IETF RFC1890, "RTP Profiles," 1996.

[36]   IETF RFC2543, "Session Initiation Protocol (SIP)," 1999.

[37]   IETF RFC2327, "Session Description Protocol (SDP)," 1997.

[38]   Third Generation Partnership Project, TS 23.228, "IP Multimedia (IM)—Stage 2," http://www.3gpp.org.

[39]   IETF RFC793, "Transmission Control Protocol (TCP)," 1981.

[40]   IETF RFC1323, "TCP Extensions," 1992.

[41]   IETF RFC896, "Congestion Control for TCP/IP," 1984.

[42]   IETF RFC1144, "TCP Header Compression," 1990.

[43]   IETF RFC2507, "IP Header Compression," 1999.

[44]   IETF RFC2508, "RTP Header Compression," 1999.

[45]   IETF RFC3095, "Robust Header Compression (ROHC)," 2001.

[46]   IETF RFC3015, "Media Gateway Control Protocol (Megaco)," 2000.

[47]   IETF RFC3031, "MPLS Specification," 2001.

[48]   IETF RFC3036, "Label Distribution Protocol (LDP) Specification," 2001.

[49]   IETF RFC2702, "Requirements for Traffic Engineering over MPLS," 1999.

[50]   IETF RFC2998, "Constraint-Based LSP Setup Using LDP," 2002.

[51]   IETF RFC2044, "UTF-8 Coding," 1996.

[52]   ETSI TS 126 234, "End-to-End Transparent Streaming Service; Protocols, and Codecs," V.4.1.0, 2001, http://www.etsi.org.

[53]   IETF RFC2326, "Real Time Streaming Protocol (RTSP)," 1997.

[54]   IETF RFC2002, "IP Mobility Support," 1996.

[55]   ITU H.323, http://www.h323forum.org/papers.

8

Core Network, Gateways, and Management

8.1 UMTS Core Interfaces and Gateways

8.1.1 Architecture

Chapter 3 of this book described the interfaces and transmission over the radio link, while this chapter covers the interfaces between the RNC and the core network together with gateways to other remote networks. The core network consists of a CS domain, a PS domain, and the Internet multimedia (IM) core. The IM was introduced in Release 5 [1] of the 3GPP specifications to cover aspects of the core network required to support IP multimedia services (especially with IPv6), and it makes use of the PS domain with predominantly IETF protocols. The purpose of the IM is to support both CS and PS applications over a single core network to the requisite standards. The interfaces to the core for CS and PS traffic are those labeled I_uCS and I_uPS in the architecture diagrams of Figures 8.1 and 8.2.

The functionality required at these interfaces is that of the RANAP [2], which includes the following main control actions:

- Management of radio access bearer setup, modification, and clear down;
- Relocation comprising soft SRNS relocations without interruption to the user, and hard inter-RNS handovers when the user is at the border of the target cell;

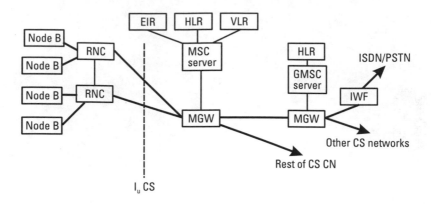

Figure 8.1 CS architecture.

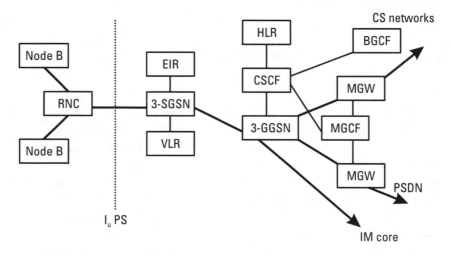

Figure 8.2 PS domain architecture.

- Release of all resources;
- Paging;
- ID management;
- Security mode control;
- Management of overload.

In addition to the control functions, the RANAP is also responsible for transferring the user traffic. These overall functions are applied to the CS and PS interfaces separately, although a common set of radio bearers is used for the two.

8.1.2 CS

The main components of the CS domain consist of two groups: parts associated with location services and mobility management that are shared with the PS-domain, and those concerned with call control and transmission for CS services. The home subscriber server (HSS) provides the master database for users, and depending on size and structure, there can be more than one HSS in a network. It is responsible for identification, numbering, addresses, security, registration, location information, subscription details, and profiles for the users. The home location register (HLR) is a subset of the HSS that contains information about the users required by the MSC in the CS domain and by the SGSN and GGSN for the PS domain. There is also a visitor location register (VLR) to hold identification and addressing details of mobiles roaming the area served by the VLR, and required by its MSCs and SGSNs.

The components of the CS domain responsible for call control and transmission are as follows:

- *Mobile services switching center (MSC):* The MSC provides the interface between the radio access network and the core network for CS services. Each MSC supports multiple RNC for UMTS services and BSS for GSM and GPRS. Its functionality is split into those of an MSC server and media gateway functions that can be either separate physical entities or a single device. The MSC is integrated with a VLR.

- *MSC server:* This is responsible for handling the call signaling, and it translates between user-network access signaling and network-network signaling for the core.

- *CS media gateway (CS-MGW):* This is responsible for handling the user traffic. It is involved in resource control through use of echo cancellers and codecs with interaction with the MSC server, media gateway control function, and gateway MSC. It may terminate bearer channels from the CS network and media streams from PS networks and provide media conversion (e.g., transcoding) where required. Where MSC server and MGW are separate, the communication between them is based on H.248/Megaco procedures.

- *Gateway MSC (GMSC):* This is an MSC that is able to take a call from a PLMN that is unable to interrogate the HLR, perform this function, and route the call onto the MSC where the mobile is located. Each MSC can be a GMSC, or else just a subset. GMSC functionality can be split between server and media gateway as for the basic MSC. Access to external CS networks is provided via a GMSC.

- *Interworking function (IWF):* This may be required to perform protocol conversion to fixed networks such as ISDN and PSTN that use other signaling systems, such as Q.931 and AC15 dual tone multiple frequency (DTMF).

- *Media gateway control function (MGCF):* This may be used to control multiple media gateways by means of H.248/Megaco procedures. MGCF functionality is included in the MSC server.

The RNC is responsible for providing the radio resources, such as RAB, required for a connection to the CS domain, and it uses the RANAP procedures [3, 4] for this purpose. The MSC server is responsible for network bearer selection following its selection of the MGW to support the bearer.

The RANAP functions make use of different protocol stacks according to whether they belong to the control-plane or the user-plane (Figure 8.3). The stack uses a set of SS7 signaling protocols to control the setup, maintenance, and release of ATM connections between the RNC and a core network node. The nature of the individual protocols involved was described in the previous chapter, as was the user-plane protocol stack. UMTS Stage 3 also uses the option of bearer independent call control [5] over ATM and IP networks, based on ITU Q.1950.1 to Q.1950.6 standards.

8.1.3 PS

The principal components of the PS domain are the SGSN and GGSN. The SGSN is responsible for providing the interface from the radio access network to the PS core, and for handling subscription and location information. Functionality of the SGSN has evolved via two stages, the 2-SGSN for GSM access networks (described in Chapter 5) and the 3-SGSN with support for UTRAN.

PS gateways are similarly based on the 2-GGSN for GPRS/GSM and 3-GGSN for UMTS. Support for the IM only arises on the timescale for 3-GSNs.

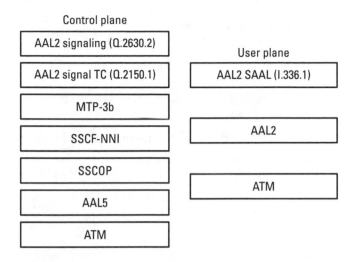

Figure 8.3 CS protocol stacks. (*After:* [3] © ETSI 2001.)

PS Gateways

The main responsibilities of the 2-GGSN are as follows:

- Provide connectivity to external PS data networks;

- Put location information into SS7 GSM MAP and send it to the HLR directly or else send the raw data via a GSM MAP converter;

- Send accounting information to a charging gateway;

- Provide connectivity to SGSNs of other PLMNs.

Other functions include provision of dynamic addresses by DHCP, DNS, and address translation, and firewall capability. Both SGSN and GGSN are implemented via IP routing switches with additional software for UMTS functions. Connectivity to the external networks is based on the creation and maintenance of PDP contexts for the relevant MS and data network. PS data for PDP contexts for an MS are encapsulated in GTP [6] for transport between the MS and GGSN via the SGSN backbone network. A GTP tunnel has two tunnel endpoint identifiers (TEID) corresponding to the SGSN for the MS and the GGSN. IP is used as the protocol for the PLMN, and GTP encapsulates the data packets for the PDP context in UDP. There is a special GTP header consisting of at least 8 bytes that is used both for control-plane messages (GTP-C) and for the user-plane (GTP-U), whose structure is shown in Figure 8.4 and described as follows:

- The initial GTP version is 1;
- The protocol type (PT) is 1 for GPRS GTP as in [6], but 0 for an earlier GTP in GSM, and * is set to zero and ignored;
- An extension header flag E is set if there is a GTP header extension;
- The flag S is set if sequence numbers (for coordination of signaling messages and responses) are included in the GTP header;
- The N-PDU number flag PN is set if the header for GTP-U contains an N-PDU number field (for use with some acknowledged mode messages);
- The message type indicates one of the 50 or so possible types so far defined (e.g., error indication, create PDP context request, forward relocation request, etc.);
- The payload length is the length in octets of the data following the standard header and including any extension headers or optional fields;
- The TEID is that of the receiving end—subsequent fields are optional.

The length of any extension headers is variable and is specified in its first octet. One example is a 2-octet PDCP sequence number for unacknowledged N-PDUs at SRNS relocation.

Vers PT * E S PN	Message type
Length	
Tunnel endpoint identifier (1st 2 octets)	
Tunnel endpoint identifier (2nd 2 octets)	
Sequence number	
N-PDU number	Next extension header type

Figure 8.4 GTP header structure (*Source:* [6] © ETSI 2001.)

The GGSN selects the appropriate external network for outgoing messages on the basis of the APN in the PDP context (see Chapter 5) and sets the APN in the PDP context for incoming messages according to the source network. The GGSN is responsible for providing the MS with a dynamic IP address (if the IP address in an MS PDP context request is blank) using either Dynamic Host Control Protocol (DHCP) or RADIUS, and it acts as the client in these cases.

3-SGSN and 3-GGSN

The 3-GSNs continue to provide the 2-GSN functions above, but also provide the UTRAN interfaces via the RNC and (from Release 5 functionality onwards) add features needed for the support of the IM over an IPv6 core network. The architecture of the PS domain with support for the IM is shown in Figure 8.2.

In this diagram, heavy lines indicate flows of user traffic, while light lines indicate administrative or signaling traffic. The main devices that complement the SGSN and GGSN for the IM are listed below:

- *Call-state control function (CSCF):* This is responsible for handling the extended Session Initiation Protocol (SIP+) for the IM (SIP+ differs from SIP [7] by addition of new message types that are required for UMTS [1]). The UE usually obtains the name of a P-CSCF and a DNS to resolve the address by DHCP after doing a GPRS attach, or directly within the PDP context activation. The UE then contacts that P-CSCF to register. UMTS uses both a serving CSCF (S-CSCF), situated on the user's home network, and a proxy CSCF (P-CSCF) on the visited network when roaming. The P-CSCF passes control of the SIP session to the S-CSCF after responding to the UE, but the interface may be internal as a single device can contain both functions. The S-CSCF will also interface to a SIP application server. One of the functions of the S-CSCF is to find the address of an interrogating CSCF (I-CSCF) on behalf of the originating UE that is the contact point for SIP sessions to a user in another network.

- *Breakout gateway control function (BGCF):* This is used if breakout is required to PSTN from the IM core. It receives requests from the S-CSCF for PSTN breakout and selects the point for breakout and relevant network and MGCF for this.

- *MGCF:* MGWs are required whenever media streams have to be terminated or translated—PSTN breakout is one such case. The MGCF controls all of the MGW.

Control of QoS is handled separately from session control in the IM. QoS signaling is performed on an end-to-end basis to negotiate or modify requirements and allocate resources, and entails both a UMTS bearer service manager for the UTRAN and an IP bearer service manager for the IP bearers. The GGSN handles the interface between these two functions, with the UE also contributing to the UTRAN procedures.

The procedures of the 3-SGSN and 3-GGSN differ slightly from those of the 2-GSNs even when performing similar functions. One such area is in PDP context activation, where for UMTS the assignment of a RAB is performed by the SGSN on receipt of the activation request. The SGSN does this by sending an RAB assignment request to the RNC that then sets up a radio bearer to the UE before returning the RAB assignment response that gives the RAB identity to the SGSN. The choice of RAB and RB depends on the QoS parameters included in the PDP context and specified to the RNC through the RAB parameters IE [2] in the assignment request. These parameters are as follows:

- QoS traffic class: conversational/streaming/interactive/background;
- RAB asymmetry indicator: symmetric bidirectional, asymmetric bidirectional, asymmetric UL/DL only;
- Maximum bit rate for (1) DL, (2) UL;
- Guaranteed bit rate (if any) for (1) DL, (2) UL;
- Delivery order (whether required or not);
- Maximum SDU size (bits) for any RAB subflow;
- For each subflow in turn, the SDU error ratio, residual BER, delivery/nondelivery of errored SDUs, and exact SDU sizes for each;
- The 95th percentile transfer delay (ms);
- Handling priority (1–15) where 1 is highest and 15 means not set;
- Allocation/retention priority when radio resources are scarce—priority level (1–15) as regards urgency for setup, degree to which it can preempt other RAB, degree to which it can be preempted;
- Source statistics descriptor (e.g., speech) to indicate statistical usage;

- Relocation requirement: real-time/lossless/none, if RAB relocated to another SRNS.

QoS is negotiable so there is also an alternative RAB parameter values IE to use if required, based primarily on alternative bit rates. For the IM there may be a special PDP context for the purpose of QoS signaling that has its own special QoS needs.

The user-plane protocol over the I_u interface between CN and UTRAN operates in one of two modes [4]: transparent mode (TrM) or support mode for predefined SDU size (SMpSDU). The choice of which to use is made by the CN at RAB establishment. TrM is used for those RABs where the traffic can be transferred over the I_u without the need for peer-to-peer protocol information exchange (e.g., GTP-U PDUs). SMpSDU is used where the traffic uses specific sizes of SDU that require some additional procedural controls (e.g., AMR speech with its multiple classes of SDU). Some of the additional procedures in SMpSDU are initialization, rate control, time alignment, frame quality classification, and error handling. These features are related to the QoS needs of different subflows over the RAB.

The CN decides which combinations of RAB subflows can be used for simultaneous transmission across the I_u interface and indicates this via the RAB subflow combination indicator (RFCI) present in each user-plane frame over the I_u. Initialization entails the signaling of each SDU size for the subflows in the RFCI. Rate control limits the maximum rate on the DL based on the guaranteed and maximum rates in the RFCI and is performed by the SRNC. The SRNC must allow the guaranteed rates but can vary the degree to which they can exceed it up to the possible maximum. The frame quality classification is required in relation to the delivery/nondelivery of erroneous SDUs and entails introduction of a CRC field at level 1.

I_u frame structures differ according to the mode. Transparent mode SDUs are variable length frames without any explicit length indication at I_u frame level, while SMpSDU uses three different frame types as follows:

- PDU type 0 is used for SDUs of defined length that require error detection. There is a fixed length header that includes the PDU type, a frame quality classification (as required for erroneous SDU delivery behavior), and the RFCI. This is followed by separate fixed length CRC fields for both the header and the complete payload, followed in turn by the variable length (but specified in RFCI) payload components.

- PDU type 1 is used for SDUs of defined length that do not require error detection. Its structure is similar to type 0 but lacks a CRC field for the payload.

- PDU type 14 is used for control procedures, such as initialization, positive, and negative acknowledgments. The fixed length header includes PDU type and procedure indication, followed by CRCs for the header and payloads. The format of the payload components depends on the procedure.

QoS Control for the IM

Network operators have the choice of whether to provide subscription-based QoS, service-based QoS, or a combination for the IM. The subscription-based approach follows the PDP context activation procedures outlined above, but for the service-based option the CSCF is also involved. In each case the IP bearer service manager is involved. This function is mandatory for the GGSN but optional for the UE. It is mandatory for the manager in the GGSN to support IP DiffServ and COPS PEP functions (see Chapter 7 and [8]), while support for IP IntServ/RSVP is optional. Where policy control is implemented, it is performed by a policy control function (PCF) within the P-CSCF, which acts as a COPS PDP, authorizing a set of resources for the session and passing these in a token (defined in an extension to SIP) to the GGSN for enforcement. The authorization includes details on IP addresses permitted and the flow, possibly using the RSVP flowspec and filter spec (see [9] and Section 7.4.3) for these purposes where supported, else DSCP flows. Support of RSVP is needed if stringent quality of service (e.g., for conversational traffic) is to be provided. This is probably most easily achieved by the use of MPLS and RSVP LSP tunnels (see Section 7.4.13 and [10]). The flows that are authorized by the PCF are based on the SDP description of the media, attributes, and bandwidth stated in the SIP INVITE (see Section 7.4.8 and [7]) for session initiation. Separate gates are defined for each direction of the session using different flows.

The actual allocation of resources is initiated by the UE following authorization of the session by use of the PDP context activation procedures.

UMTS Gateways

In order for UMTS users to be able to access a full range of services, they will be required to communicate with devices that use other sets of standards, and to do so will need specific media gateways. In general these gateways will all be controlled by the MGCF, and this in turn will interact with the S-CSCF for authorization of media flows and resource allocation. The simplest of these are the transcoders to convert from one speech codec to another (see Section 10.1 for codecs), while another simple type might be to convert between different MPEG-4 formats (see Section 10.2) for the benefit of simple handsets, but others are potentially much more complicated.

One of the main complex examples is access by 3G terminals to H.323 (PS) multimedia sessions (see Section 7.5). It is not mandatory for 3G providers to offer this service, but as H.323 is the main standard for PS multimedia, there is likely to be a significant demand for it. Two forms of conversion are required for basic peer-to-peer communication. The first is translation between session control messages based on SIP in UMTS and the H.225 and H.245 messages used in H.323. This function entails address conversion, the mapping of messages from one protocol to another, and also the conversion from SDP descriptions in SIP of terminal requirements and capabilities in UMTS to H.245 capability sets for the remote H.323 terminals. The range of functions for the gateway (or interworking function) to perform also depends on whether SIP servers and H.323 gatekeepers are present, or if the gateway itself is to perform these roles. SIP translation is not straightforward, as a gateway has to know the context of any message before it can convert it to the correct ISUP (for ATM core), H.225, or H.245 message, since general service messages in both SIP and ISUP can be used in a variety of ways. At the most basic level there is a correspondence between SIP INVITE and H.323 SETUP (or ARQ if there is an H.323 gatekeeper), between SIP 200 OK and a wide range of H.323 successful operation messages, between SIP 400/500 series and H.323 failure messages, and between SIP BYE and H.245 end session/H.225 release complete.

A more serious problem is the difference in style and capability of SIP/SDP as compared to H.225/H.245. SIP/SDP essentially advertises the characteristics of the media, including a single RTP/UDP port, to be used for a session, whereas H.225 initiates the session, with H.245 being used to negotiate the media features to be used, and including separate RTP and RTCP sessions. Unless the H.245 fast connect option introduced in H.323v2 is used, there is no way for the originating SIP gateway to know which media description to use. In the case of fast connect, the usual H.245 capability exchange is replaced by the caller sending a list of possible media, and the receiver returning a definitive selection from the list. But even then a SIP gateway receiving an incoming fast connect still has problems, as it does not know which of the received capabilities to forward to the UE (unless extended to pass all), except perhaps by using CSCF functions to check the destination UE's own capabilities and subscription rights. One solution to this might be a SIP options exchange of SDP media descriptions prior to any SIP invitation. SIP has fewer logical channel procedures than H.245, leading to further difficulties (see [11] for a more detailed discussion of these issues). Once session control has been established, the gateway would also have to perform transcoding for various incompatible codec sets for both audio and video functions, unless the H.323 range is extended. 3G-324M mandates use of UMTS AMR voice codecs or optionally G.723 (as in the original H.324 recommendation), while H.323 (up to v.5) excludes AMR from its list of audio codecs. So, unless both the UMTS and H323 terminals are using G.723 voice, transcoding is necessary. At the video level, only H.261 is mandatory for H.323 multimedia support, but H.263 is optional for both UMTS and H.323 and can avoid transcoding here. The media gateway must also convert between SCTP for UMTS IP transport and RTP/UDP for H.323 media streams and TCP for H.323 signaling. Detailed interworking specifications will be required to handle all these issues and will entail the extended SIP+ [1] rather than the original SIP.

In order to support more than just the simple peer-to-peer communication, the media gateway needs to perform H.323 gatekeeper and multipoint control unit functions also.

8.2   cdma2000 Core Interfaces and Gateways

8.2.1   Architecture

The architecture for cdma2000 shown in Figure 8.5 is simpler than that for UMTS and based on either CS for speech or Mobility IP for packet data and multimedia. The mobiles in Figure 8.5 have radio links to a BTS that is controlled by the BSC, which communicates with the MSC for CS traffic and with a router with IWF for IP traffic, with authentication and authorization provided by the RADIUS server [12]. Heavy lines indicate paths used for user traffic, while light lines are only for signaling and administration. A2 and A5 are the interface reference points between BSC and MSC for user CS PCM and byte-stream traffic, respectively, and A10 that between the PCF function of the BSC and the router for user PS traffic [13]. A1 handles signaling and administration to the MSC, while A11 represents signaling from PCF to router.

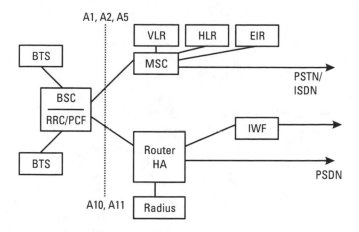

Figure 8.5 Cdma2000 architecture.

8.2.2 CS

The CS components are as follows:

- *MSC:* This performs a similar role to the MSC in UMTS, but uses a different standard, IS-634, for communication with the BSC. It has the ability to perform both transcoding and tandeming when required.
- *HLR/AUC:* This too has a similar role to the HLR in UMTS but stores the analogous data in a different way. The authentication center can be separate from the HLR. The mobile's identity is defined by the ESN and 10-digit mobile identification number (MIN) instead of the 15-digit IMSI, while the mobile's capability covers a different range of features to those of the UE. Communication with the HLR over the core network uses the ANSI-41 extension to SS7 instead of GSM-MAP.
- *VLR:* The visitor location center is analogous to that for UMTS but holds information in a different format.
- *EIR:* Release B of cdma2000 includes an equipment identification register, but its role was not defined at that stage.

As with UMTS, SS7 signaling techniques may be used between the BSC and MSC to set up CS or ATM connections for calls. Again, this uses MTP [14] and SCCP [15] aspects, but it is not a full implementation [13]. Where ATM is used, signaling is carried in AAL5 cells, while user traffic is carried in AAL2.

MTP3 is simplified mainly on account of there being only one signaling route between the BSC and MSC, so that most rerouting features do not apply. Another simplification is that the BSC automatically rejects messages with incorrect destination point codes, and signaling traffic flow control is not applicable.

The main omissions from the full SCCP specification [13] are error detection, acknowledgment, and flow control [14], while segmentation and reassembly are to be used if the signaling message exceeds the maximum for MTP. There is one signaling connection for each active MS. The part of SCCP that is used is the base station application part (BSAP) that consists of two components: the direct transfer application part (DTAP) and the base station management application part (BSMAP), of which DTAP is the section relevant to call control and quality. The main DTAP messages for call control are the service request and paging request. Each of these messages includes detailed MS IS-2000 capabilities and the SO that is required. The most important speech SOs are 13K speech [rate set 1 (RS1); see Section 4.4.6] and EVRC [rate set 2 (RS2)], while data includes G3 FAX, ISDN, Async data for RS1 and RS2, and 3G high-speed packet data. The MSC uses BSMAP messages to tell the BSC to assign or releases the required resources for these services. The BSMAP assignment message for PS calls contains QoS parameters. For unassured mode these are the user's subscribed values from the HLR, but in Release A the only QoS parameter used was the radio priority.

8.2.3 PS

The PS core network is based on Mobility IP implemented on standard IP routing switches, but with the addition of a RADIUS server [16, 17] and an IWF. Mobile access can be based on either simple IP or Mobility IP, and uses a PPP session between the mobile and the PDSN in either case, but with differing procedures across the R-P interface in Figure 8.5. Base station control within the radio network is based on RRC and packet control functions (PCFs), with the latter responsible for the R-P interface. The core PDSN components are as follows:

- *Router:* cdma2000 aims to use existing standards wherever possible, so the router is simply a standard IP router, but with the otherwise optional IP Mobility (see Section 7.4.15 and [18]) software modules. The main features of this are home agent and foreign agent capability.

- *IWF:* The role of the interworking function is to perform translation between formats for the PSDN and fixed networks, such as ISDN, with at least one interface for each such network. There are a number of other functions that may be in either the IWF or the routing switch; these include SIP proxy server support, media gateways, and MGCF. RTP mixer functionality is also desirable to support mobiles involved in multicast multimedia sessions. Some of these functions may be included in the router itself.

- *Remote authentication dial-in user service (RADIUS):* The RADIUS server performs accounting, authentication, and authorization (AAA) roles for PS services. The PDSN uses both home and local RADIUS servers but employs them differently for simple IP and Mobility IP. In simple IP the PDSN acts as a RADIUS client and sends the mobile user's Password Authentication Protocol (PAP) or Challenge Handshake Authentication Protocol (CHAP) authentication information for the PPP session to the local RADIUS server in a RADIUS-ACCESS request. This includes the user name in the form @realm, where MSID is the mobile station ID and realm is the Internet domain of the PSDN, followed by an IP address for the PSDN and the PAP/CHAP fields. The MSID can be any of IMSI, MIN, or international roaming number (IRN). The RADIUS server returns an accept message if the data is valid, and data transfer can commence. The PPP session is maintained even if the mobile is dormant, and the PSDN sends accounting information to the RADIUS server at the end. For Mobility IP, session control is based mainly on Mobility IP home and foreign agent (HA/FA) registration and challenge procedures [19] at PPP session start instead of PAP/CHAP, but with user foreign agent challenge (FAC) information sent to the RADIUS server by the PSDN on receipt of the Mobile IP registration request. Optional IP security between PSDN and RADIUS server and between PSDN and the home agent may be provided using IPSec and Internet key exchange (IKE); if so, the RADIUS server controls key distribution and which, if any, security procedures to use.

8.3   Performance Network Management

8.3.1   UMTS

Network management consists of two primary functions: administrative and performance management. The former consists of maintenance of the numerous registers (e.g., HSS, HLR, VLR, EIR) that are needed for access to the services and for mobility management. The second category covers configuration, resource allocation, network statistics, and fault control [20]. Performance management includes the operational statistics of the administrative management (i.e., the above registers, as well as those devices that handle the calls directly, such as base stations, RNC, and GSNs). This section is concerned only with performance management as it is the part that is directly related to quality of service. QoS management in UMTS is based on policy provisioning to meet the needs of user SLAs and on QoS monitoring [20]. The former uses the CSCF PCF functions mentioned in Section 8.1, while the latter depends on the collection of statistics.

UMTS suggests that statistics be gathered at regular intervals that can be any of 5, 15, 30, or 60 minutes [21] as seems most appropriate for the individual statistic. Reports can either be submitted regularly by the network element being monitored or be requested periodically by the relevant element manager (e.g., the Node B in UMTS makes regular reports to the RNC that can in turn be processed and sent to an overall network management center). A rough outline of the main types of statistic collected per device category is listed below.

RNC

- Number of successful requests for packet and CS services;
- Number of unsuccessful requests subdivided by failure causes;
- Average interval between requests for CS services and PS services separately;
- Number of successful paging messages;
- Number of unsuccessful paging messages by failure cause;
- Number of successful internal handovers between channels;
- Number of unsuccessful internal handovers by failure cause;
- Number of successful soft handovers to other RNC;
- Number of unsuccessful handovers to other RNC by failure cause;
- Number of hard handovers from UMTS to GSM or vice versa and failures;
- Numbers of hard handovers to and from other mobile networks;
- Amount of data discarded.

Node B or BTS

- Numbers of soft and softer handovers and failure rates and causes;

- Numbers of hard handovers to GSM and other services and failure rates;
- Numbers of paging messages lost;
- Channel utilization statistics;
- Number of packet channel assignment requests, accepts, and rejects;
- Number of service upgrades/downgrades.

MSC

- Numbers of attempted, successful, and failed incoming and outgoing CS calls;
- Duration and bandwidth of calls;
- Numbers of attempted, successful, and failed cipher mode commands;
- Numbers and types of attempted, successful, and failed HLR interrogations;
- Numbers of EIR interrogations and outcomes;
- Statistics on IMSI attach/detach;
- Use of TMSI identification;
- Inter- and intra-MSC handover statistics;
- MGCF statistics.

HLR

- Numbers of roaming mobiles outside HPLMN;
- Numbers of subscriber updates;
- Numbers of location updates.

VLR

- Number of visitors arriving;
- Number of location updates;
- Number of authentications in VLR and requests to HLR;
- MS short message availability messages to HLR;
- Identification and paging request statistics.

EIR

- Numbers of IMEI identification checks and outcomes.

SGSN

- Numbers of frames sent and received over GSM LLC for AM and UM, plus AM retransmissions;
- Numbers and volumes of GSM SNDCP PDUs sent and received;
- Numbers and volumes of UMTS RLC and MAC PDUs sent and received for TM, AM, and UM;
- Numbers of errors;
- Statistics for BSSGP traffic;
- Numbers of successful/unsuccessful IMSI and GPRS attach/detach;
- Numbers of routing and location updates;
- Numbers of HLR and EIR transactions;
- Numbers of subscribers in each of GPRS standby and ready states;
- Numbers of successful/unsuccessful incoming and outgoing PDP context activations and deactivations;
- Charging information for radio link use.

GGSN

- Statistics on PDP activations/deactivations per APN;
- Charging information for external network use.

CSCF

- Statistics on registrations per P-CSCF and S-CSCF;
- SIP session statistics.

BGCF

- Numbers of PSTN breakouts from IM.

UE

- Measures of error rates, signal strengths, and round-trip times.

Most of this data affects QoS indirectly through detecting areas of potential overload of network components, so that their capacity can be increased before serious congestion results. The measurements by the UE and RNC are vital for maintenance of signal quality over very short periods through the use of outer loop power control (see Chapter 3).

The MSC and SGSN are responsible for sending the data required for billing of CS and PS calls to the charging facility. The main essentials are the duration and bandwidth of CS calls and the number of packets sent for PS, but the full QoS profile is also needed if the network operator's charges are also based on quality.

In addition to these general statistics relevant to operation of the mobile network, there is also a need for proprietary fault management messages for the various types of equipment used in the network. Typically, this will consist of alarm messages, graded by severity and type, for component failures. In the core network these will often be based on IP SNMP and sent to an appropriate management platform, but the general approach to fault management in UMTS is based on the ITU's Common Management Information Protocol (CMIP) and X.700 series of recommendations instead [22]. A single component failure will usually cause multiple alarms; so to simplify the response, there should be some automatic fault correlation modules to determine the root cause of a series of related messages and to invoke automatic scripted corrective action where possible. The SGSN and GGSN in the core network are based on upgraded IP routing switches, and so support standard features for router management information bases (MIBs) and RMON2 traffic statistics in addition to the specific requirements of mobile radio. This applies equally to the routing switches used to support Mobility IP in cdma2000 networks. ATM switches for the CS core also support standard ATM management procedures in addition to the radio requirements.

Alarms can also be generated for the crossing of configurable performance thresholds using integration reference point (IRP) recommendations for UMTS fault management [22]. Installation, upgrades, and expansion of the UTRAN also require specific network management procedures, and those for the RNC and Node B are given in [23].

8.3.2 cdma2000

Overall management principles are the same for cdma2000 as for UMTS, but in cdma2000 everything is based on use of SNMP rather than CMIP, and administrative calls to HLR/VLR use ANSI-41 instead of GSM-MAP.

Call records for billing of PS calls are sent to the RADIUS accounting facility.

References

[1] Third Generation Partnership Project, TS 23.228, "IP Multimedia (IM)—Stage 2," http://www.3gpp.org.

[2] ETSI TS 125 413, "UTRAN I_u Interface RANAP Signaling," V.4.2.0 (2001), http://www.etsi.org.

[3] ETSI TS 125 410, "UTRAN I_u General Aspects and Principles," V.4.2.0 (2001), http://www.etsi.org.

[4] ETSI TS 125 415, "UTRAN I_u Interface: CN-UTRAN User-Plane Protocols," V.4.2.0 (2001), http://www.etsi.org.

[5] ETSI TS 129 232, "Media Gateway Controller—Media Gateway Interface," V.4.2.0 (2001), http://www.etsi.org.

[6] ETSI TS 129 060, "GPRS Tunneling Protocol (GTP)," V.4.2.0 (2001), http://www.etsi.org.

[7] IETF RFC2543, "Session Initiation Protocol (SIP)," 1999.

[8] IETF RFC2753, "Framework for Policy-Based Admission Control," 2000.

[9] IETF RFC2205, "Resource Reservation Protocol (RSVP)," 1997.

[10] IETF RFC3209, "RSVP-TE: Extensions to RSVP for LSP Tunnels," 2001.

[11] Stephens, A., and P. J. Cordell, "SIP and H.323—Internetworking VoIP Networks," *British Telecom Tech. J.*, Vol. 19, No. 2, 2001, pp. 119–127.

[12] 3GPP2 P.S0001-A, "Wireless IP Network Standard," http://www.3GPP2.org.

[13] 3GPP2 A.S0001, "3GPP2 Access Network Interfaces Interoperability Specification," http://www.3gpp2.org.

[14] ANSI T.1.111, "Signaling System Seven (SS7)—Message Transfer Part (MTP)," 1992.

[15] ANSI T.1.111, "Signaling System Seven (SS7)—Signaling Connection Control Part (SCCP)," 1992.

[16] IETF RFC2138, "Remote Authentication Dial-In User Service (RADIUS)," 1997.

[17] IETF RFC2139, "RADIUS Accounting," 1997.

[18] IETF RFC2002, "IP Mobility Support," 1996.

[19] IETF RFC3012, "MobileIP AAA," 2000.

[20] ETSI TS 132 101, "3G Telecom Management: Principles and High-Level Requirements," V.4.2.0 (2001), http://www.etsi.org.

[21] ETSI TS 132 104, "UMTS Performance Management," V.3.4.0 (2001), http://www.etsi.org.

[22] ETSI TS 132 111-1/2/3/4, "UMTS Fault Management," V.4.0.0 (2001), http://www.etsi.org.

[23] ETSI TR 132 800, "Management Level Procedures," V.4.0.0 (2001), http://www.etsi.org.

9

UMTS Classes of Service

9.1 Basic Classes

UMTS and GPRS both define QoS classes [1], but cdma2000, at least in its early formulations, does not. Users of these services may communicate with both fixed networks and other mobiles, so end-to-end performance is also influenced by the features of remote networks on which other parties may be situated. Although the capabilities of UMTS and GPRS differ widely, they both support the same four QoS classes:

1. Conversational;
2. Streaming;
3. Interactive;
4. Background.

The characteristics of each of these classes are described in the following sections.

9.1.1 Conversational Class

As its name implies, this is the class to which conversations belong. It applies to not only speech itself, but to any application that involves person-to-person communication in real-time, such as videoconferencing and interactive video games.

The basic qualities required for speech itself are low delay, low jitter (delay-variation), reasonable clarity (codec quality), and absence of echo. In the case of multimedia applications, such as videoconferencing, it is also necessary to maintain correct relative timing of the different media streams, for example, lip synchronization.

This class is tolerant of some errors, as dropping or corruption of a voice packet lasting for a typical 20 ms is unlikely to be detected by a user. The degree of error protection required is variable with some applications. Some compression algorithms divide the subject material into different sets whose importance differs. For example, with H.324 videoconferencing, information associated with lip movement is more critical to quality than background scenery, and hence may be offered a higher level of error protection; while for speech, the AMR codec for UMTS has three distinct categories of information bits.

One of the key preliminaries to QoS negotiation for this class is the choice of codec to use. If the codecs used by the two peer end users were different, then there would be a need for transcoding by a media gateway leading to major delays and probable loss of content quality. For CS and ATM core networks, this involves H.245 capability exchanges; while for IM core networks and remote IP networks, the initial SIP INVITE contains an SDP list of available codecs for the recipient to respond to, with the final decision made by the session initiator. The choice of codec must also be compatible with the network resources available, and so is influenced by RNC, SGSN, and GGSN QoS profile negotiations. The role of gateways and control functions for this was described in Section 8.1.3.

9.1.2 Streaming Class

The streaming class consists of real-time applications that send information to a viewer or listener, but without having any human response. Examples of this include video-on-demand, live MP3 listening, Web-radio, news streams, and multicasts.

Because of the absence of interaction, there is no longer a need for low delay, but the requirements for low jitter and media synchronization remain. The error tolerance remains, but higher quality is required when listening to hi-fi music than for speech.

The removal of the low delay criterion makes it possible to use buffering techniques in the end-user equipment to even out the delay variation, so the acceptable level of network jitter is higher than for the conversational class. As in the conversational case, the choice of codec needs to be negotiated.

9.1.3   Interactive Class

This class covers both humans and machines that request data from another device. Examples of this include some games, network management systems polling for statistics, and people actively Web-browsing or searching databases.

The first requirement is for a delay that is within an application time-out, or reasonably prompt, for the human activities, but not as low as for the conversational class. The second need is for data integrity.

9.1.4   Background Class

Background class covers all applications that either receive data passively or actively request it, but without any immediate need to handle the data. Examples of this include e-mails, short message service, and file transfers.

The only requirement is for data integrity, although large file transfers will also require an adequate throughput.

9.2   UMTS QoS Implementation

9.2.1   Architecture and QoS Profile

The basis of provision of the required QoS in UMTS is the selection of bearers with the appropriate characteristics. Several different bearers will be used in support of any call, and the different types are shown in the bearer architecture diagram of Figure 9.1.

Each UMTS bearer service is characterized by a number of quality and performance factors, and these are listed below. Some applications entail several distinct subflows over the radio access bearers that require separate QoS requirements, especially error protection, but otherwise they use the same characteristics as the UMTS bearer. Where a radio access bearer has more than one subflow, it also has a corresponding number of radio bearers.

- *Traffic class:* This is conversational, streaming, interactive, or background.
- *Maximum bit rate (Kbps):* This is the maximum number of bits that a UMTS bearer can deliver to a service access point (SAP) in a

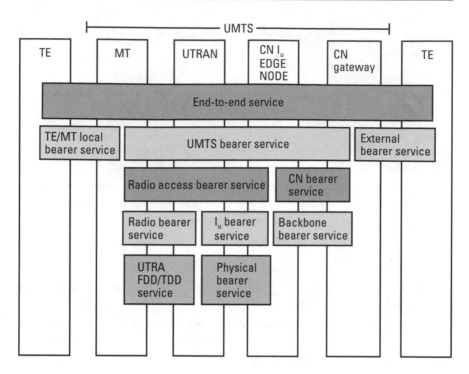

Figure 9.1 UMTS QoS architecture. (*Source:* [1] © ETSI 2002.)

specified interval. It is required in order to reserve radio resources, such as codes, on the DL. It limits the peak transient rate that can be supported, and controls selection of the appropriate peak rate for an application that can operate at a number of speeds. There is an optional token bucket formula for measuring conformity.

- *Guaranteed bit rate (Kbps):* This is the maximum number of bits that the UMTS guarantees to deliver to a SAP in a specified time. It specifies the minimum resources required and is used to support admission control. The token bucket algorithm can also be applied to this.

- *Delivery order (y/n):* This parameter determines whether the bearer sequences SDUs in the correct order or not. The choice depends on the user application's protocol rather than the class.

- *Maximum SDU size (octets):* This is used in admission control and policing.

- *SDU format information (bits):* This is a list of possible exact SDU sizes for an application. It is needed for applications that use the RLC transparent mode and for applications that require unequal error protection for different payload components over the radio access bearer.

- *SDU error ratio:* This is the fraction of SDUs that are either lost or detected as erroneous for traffic that conforms to its agreed contract. It is mainly used internally in the UTRAN to configure protocols, algorithms, and error detection schemes. Different subflows over the radio access bearer may have different requirements.

- *Residual bit error rate:* This indicates the undetected bit error rate in delivered SDUs if error detection is used, or the total bit error rate if not. It is used to configure protocols, algorithms, and error detection codes. BER is specified for each subflow over the radio access bearer.

- *Delivery of erroneous SDUs (y/n/–):* This indicates whether erroneous SDUs should be delivered or dropped, or delivered without considering error detection. Again, this depends on subflow.

- *Transfer delay (ms):* This indicates the maximum delay for the 95th percentile of the distribution of delay for all delivered SDUs during the lifetime of a bearer service, where delay for an SDU is defined as the time from a request to transfer an SDU at one SAP to its delivery at the other SAP. It is not meaningful for bursts of multiple SDUs that may get queued behind each other, and in such cases only applies to the start of the burst. This application delay tolerance is used to set transport formats and ARQ parameters.

- *Traffic handling priority:* This determines the relative importance of handling all the SDUs on one bearer as compared to another. It is primarily used for scheduling different types of interactive traffic.

- *Allocation/retention priority:* This is used to discriminate between bearers when allocating or retaining scarce resources. It is a subscription attribute rather than something that can be negotiated by the mobile.

- *Source statistics descriptor (speech/unknown):* This indicates whether the source has a statistical character with discontinuous transmission characteristics like those of speech that permit statistical multiplexing to be employed.

The applicability of each of these bearer properties is summarized in Table 9.1.

The main differences are in bit rate and the exact way in which the attribute is used. On the RAB header, compression will often be used, which affects the guaranteed bit rate, but not the maximum, which has to allow for some uncompressed traffic. The value of the transfer delay for the RAB is less than that for the UMTS bearer as it forms one contribution to the overall value specified by the latter.

The residual BER and SDU error ratio should also be less for the RAB than the UMTS due to contributions from the core. The SDU Format Indicator and Source Statistics Descriptor are only relevant to the RAB. IP-based I_u bearer services are required to use DiffServ for QoS control, and this is also to be used for interaction with ATM SVCs. The actual mapping from UMTS QoS classes to DiffServ code-points is left to the operator. Core bearer services are also required to use DiffServ both for IP and its interaction with ATM.

Table 9.1

Bearer Attributes for Each Bearer Traffic Class for UMTS Bearers (U) and Radio Access Bearers (R)

Traffic Class	Conversational Class	Streaming Class	Interactive Class	Background Class
Maximum bit rate	U, R	U, R	U, R	U, R
Delivery order	U, R	U, R	U, R	U, R
Maximum SDU size	U, R	U, R	U, R	U, R
SDU format information	U, R	U, R		
SDU error ratio	U, R	U, R	U, R	U, R
Residual bit-error ratio	U, R	U, R	U, R	U, R
Delivery of erroneous SDUs	U, R	U, R	U, R	U, R
Transfer delay	U, R	U, R		
Guaranteed bit rate	U, R	U, R		
Traffic-handling priority			U, R	
Allocation/retention priority	U, R	U, R	U, R	U, R
Source statistics descriptor	U, R	U, R		

Note: These attributes are applied in slightly different ways to the different types of bearer [i.e., UMTS bearer or RAB, I_u bearer or core bearer].
Source: [1] © ETSI 2002.

Control Functions

The manner in which these attributes are handled is governed by a set of control functions that may be subdivided into those for call control and those for user traffic. The call control functions are as follows:

- *Service manager:* This coordinates the control functions as a whole.

- *Translation:* This function converts between the UMTS operations and those of the interfacing external networks. In particular, it converts between the UMTS bearer attributes and those of the QoS requirements of the external network's service control protocol, such as the TSPEC for IP IntServ.

- *Admission/capability control:* This maintains information about all the network resources and about those for specific bearers. It checks resource availability for bearer requests and allocates and reserves resources as appropriate.

- *Subscription control:* This checks the administrative rights for the use of a bearer service and QoS attributes.

The functions for control of the user traffic are subdivided as follows:

- *Mapping function:* This marks the data units in such a way that they receive the correct QoS.

- *Classification function:* This is used where a mobile has an established session using several bearer services in order to assign each data unit to the right bearer for the QoS as determined from the unit header or other traffic characteristic.

- *Resource manager:* This distributes resources between competing services according to QoS needs by means of techniques like scheduling, bandwidth management, or power control.

- *Traffic conditioner:* This performs policing or traffic shaping to ensure that it conforms to the negotiated QoS for the service. Nonconformant traffic is dropped or else tagged to allow dropping during congestion.

In general each of the types of bearer service has its own management functions as above. There are a number of different scenarios for establishing the QoS depending on the types of network, service required, and terminal capability [2]. For the PS domain the requested QoS is put into the QoS Profile IE, whose format is shown [3] in Figure 9.2:

Some UE are unable to request a specific QoS for a service and merely request a value based on their subscription (see [4] and Section 9.2.2) by inserting zero for each of the bits of the relevant fields above in the UE-to-network direction. The main features of the other values for the individual fields are listed below.

- *Reliability:* This determines the reliability over GTP, LLC, and RLC via acknowledged or unacknowledged modes and whether the data is protected or not.

- *Delay class:* There are four classes (1–4) defined by bit sequences 001, 010, 011, and 100, respectively, where class 4 is best effort.

- *Precedence class:* The three classes (high, medium, low) are coded by bit sequences 001, 010, and 011, respectively.

- Peak throughput is given in binary multiples of 1,000 octets/s coded from 0001 (1,000 octets/s), 0010 (2,000 octets/s), up to 1001 (256,000 octets/s).

8	7	6	5	4	3	2	1	
Quality of service IEI								1
Length of quality of service IEI								2
0	0	Delay class			Reliability class			3
Peak throughput				0	Precedence class			4
0	0	0	Mean throughput					5
Traffic class			Delivery order		Delivery of erroneous SDU			6
Maximum SDU size								7
Maximum bit rate for uplink								8
Maximum bit rate for downlink								9
Residual BER				SDU error ratio				10
Transfer delay					Traffic handling priority			11
Guaranteed bit rate for uplink								12
Guaranteed bit rate for downlink								13

Figure 9.2 QoS Profile IE (*Source:* [3] © ETSI 2001.)

- Mean throughput is coded differently in multiples of 100 octets/hr from 00001 (100), 00010 (200), 00011 (500), 00100 (1,000), 00101 (2,000), 00110 (5,000), up to 10010 (50,000,000), but with 11111 indicating best effort.

- Erroneous SDU behavior has 001 for no error detection, 010 for delivery of erroneous SDUs, and 011 for nondelivery.

- The four QoS traffic classes are coded as 001 for conversational, 010 for streaming, 011 for interactive, and 100 for background.

- Maximum SDU size is coded in multiples of 10 octets up to 1,500 via values 00000001 to 10010110, while 10010111, 10011000, and 10011001 code the specific sizes 1,502, 1,510, and 1,520, respectively.

- The maximum bit rates for UL and DL both use the same format, which is as follows: values 00000001 to 00111111 define rates of 1 to 63 Kbps in steps of 1 Kbps; values 01000000 to 01111111 define rates from 64 to 568 Kbps in steps of 8 Kbps; and 10000000 to 11111110 define rates from 576 to 8,640 Kbps in steps of 64 Kbps.

- The residual BER uses bit sequences 0001 to 1001 to indicate BER from 5×10^{-2} to 6×10^{-8}.

- The SDU error ratio uses bit sequences from 0001 to 0110 to indicate ratios from 1×10^{-2} to 1×10^{-6}, with 0111 to indicate 10^{-1}.

- Traffic handling priority only applies to interactive class. Sequences 01, 10, and 11 correspond to priority levels 1, 2, and 3.

- Transfer delay bit sequences 000001 to 001111 define delays from 10 to 150 ms in steps of 10 ms; values 010000 to 011111 define delays from 200 to 950 ms in steps of 50 ms; and values 100000 to 111111 define delays from 1,000 to 4,100 ms in steps of 100 ms. This field is ignored for interactive and background traffic.

- Guaranteed bit rates for UL and DL are coded similarly to the maximum bit rates. This field is ignored for interactive and background traffic.

The manner in which the QoS IE in the PDP context activation request determines the RAB parameters was discussed in Chapter 8.

A single application may contain several distinct subflows (e.g., audio and video or subcomponents of each) that have different QoS needs. In that case the RNC allocates separate radio bearers under the RAB for each subflow (see Section 10.1.1 for the example of AMR speech). The length field in the QoS IE allows for the separate needs of multiple flows for a single application to be quoted together, subject to a maximum of 254 bytes.

9.2.2 Subscription Classes

As indicated above, subscription classes are used to determine the QoS to be provided for early GPRS users and subsequently for many basic users. This is based on three reliability classes and four delay classes [4]. The reliability classes characterize the probability of loss, duplication, sequencing error, and corruption of an SDU. Class 1, which is intended for error-sensitive applications, specifies that these should each be less than 10^{-9}, while classes 2 and 3 for applications of limited sensitivity and insensitivity, respectively, are much less demanding. Both of these specify 10^{-5} for duplication and sequencing errors, while loss and corruption probabilities of up to 10^{-2} are acceptable for the insensitive class 3 with up to 10^{-4} and 10^{-6}, respectively, for class 2.

The delay classes describe the average and 95th percentile delays for SDUs of lengths 128 and 1,024 bytes across the radio links and GPRS network, but excluding any external networks. The most demanding group is class 1, which specifies a maximum of 0.5 seconds average delay for 128-byte SDUs and 2 seconds for 1,024, with 95th percentiles of up to 1.5 and 7 seconds, respectively. These figures are much higher than the targets or conversational applications (see Table 9.3), while classes 2 and 3 quote figures that are greater by the order of factors of 10 and 50, respectively, and the remaining class 4 is purely best effort. Throughput is negotiable rather than being determined by subscription.

9.2.3 Establishing End-to-End QoS

The precise services offered over UMTS will depend on both the policy of the network operator and the capabilities of the mobile equipment. UMTS expects that an application used by the UE should be able to specify its QoS needs, but the mechanism for this is outside the UMTS specifications, although one option is the SIP/SDP procedure and another is the use of H.245. Early model UE are unlikely to be able to negotiate detailed QoS needs, so operators may initially offer QoS based purely on subscription rights (as in Section 9.2.2) or the capability IE of the UE, and retain this for any pay-as-you-go customers later on, while introducing negotiable QoS for more sophisticated users. As a result, UMTS offers several distinct scenarios for establishing end-to-end QoS and distinguishes between the mechanisms for the original ATM-based CS and PS domain core on one hand and the IM core on the other. The original method is the UMTS QoS mechanism whereby the QoS profile contained in the PDP context activation method is translated into B-ISUP QoS parameters that are included in the initial address message and ATM connection request, while the IM introduces the IP bearer service manager to handle QoS for IP. Some possible scenarios that apply to the IM [2] are given below.

1. The UE has no IP bearer service manager, although it may support UMTS QoS and include a QoS profile in the PDP context activation request. In this case, the GGSN controls the QoS towards the remote terminal, while the latter (or its own GGSN) controls that towards the UE. The GGSN will have support for DiffServ and COPS and so can interact with the P-CSCF to authorize a flow based on the SIP SDP media description, user's subscription rights, and resources actually available in the UTRAN.

2. The UE supports UMTS QoS mechanism and has an IP bearer service manager but does not support RSVP. In this case, the application QoS needs specified by SIP SDP parameters are mapped to IP QoS needs and then to the QoS profile to put in the PDP context activation. If the UE has received a SIP authorization token from the CSCF, then this also goes into the activation request for the session.

3. The UE also supports RSVP. Here the application needs are mapped to RSVP flowspec parameters and included in the QoS profile for the PDP context activation. If the UE has received an authorization token, then it includes that in both the activation request and in RSVP messages.

4. In some instances, where the UE does not support RSVP, but the GGSN does, then the latter may act as an RSVP proxy for the UE.

In all these IM instances, the P-CSCF is responsible for authorization of resources on receipt of a SIP message that contains the SDP description of the session. The P-CSCF establishes a policy control function for the session and returns an authorization token to the originating UE in a SIP 183 call progress message and the SIP INVITE to the destination UE. Once the GGSN has requested bearer authorization and received a SIP 200 OK from the far end, it enables the authorization by opening the gate in the GGSN.

9.3   UMTS QoS Targets

UMTS defines specific QoS targets to provide an adequate service to mobile users [5]. These targets depend on the mobility of the user and on the type of environment. The effect of mobility on transmission rates is that required by the IMT-2000 specification (Table 9.2).

In UMTS the maximum speed envisaged for the high mobility category is 500 km/hr using terrestrial services to cover all high-speed train services and 1,000 km/hr using satellite links for aircraft. The data-rate bands are also related to the environment and cell size in UMTS with rural environments and satellite links being restricted to 144 Kbps aggregate for a single mobile and with 2 Mbps only available as an instantaneous rate rather than a guaranteed rate (Section 6.2 indicates some of the problems on cell size associated with 2 Mbps or other high rates). The peak rate for the GERAN (see Chapter 5) and for UWCC networks is only 384 Kbps.

The targets for specific QoS classes are specified in Tables 9.3 to 9.5 [5].

The UMTS has been designed to meet these targets for devices for source and destination both located on the same UMTS network, but some of the delay variation targets can only be met where external networks are involved if the user application has a long playback buffer and precise time-stamps to eliminate the network jitter.

The lip synchronization limit of 100 ms in the first of these tables for video phones will be quite noticeable to users, as the lower limits for detectability are about 30 ms for image in advance of speech, and only 10 ms where speech precedes image (since the brain allows for the slow speed of sound). The need for minimal delays for this application prevents the use of decoders with long play-out buffers to avoid this. No target is set for streaming audio-video where this restriction does not apply.

Although access probability and call retention probability are included in the list of QoS attributes, no specific targets are set despite their importance to the user.

Table 9.2
IMT-2000 Mobility Categories

Category	Physical Speed (km/hr)	Data Rate
Limited mobility	Up to 10	2 Mbps
Full mobility	Up to 120	384 Kbps
High mobility	More than 120	144 Kbps

Table 9.3
End-User Performance Expectations—Conversational and Real-Time Services

Medium	Application	Degree of Symmetry	Data Rate (Kbps)	Key Performance Parameters and Target Values		
				End-to-End One-Way Delay (ms)	Delay Variation Within a Call (ms)	Information Loss (%)
Audio	Conversational voice	Two-way	4–25	< 150 preferred < 400 limit (Note)	< 1	< 3 FER
Video	Videophone	Two-way	32–384	< 150 preferred < 400 limit Lip-synch: < 100		< 1 FER
Data	Telemetry— two-way control	Two-way	< 28.8	< 250	—	0
Data	Interactive games	Two-way	< 1	< 250	—	0
Data	Telnet	Two-way (asymmetric)	< 1	< 250	—	0

Note: The overall one-way delay in the mobile network (from UE to PLMN border) is approximately 100 ms.
Source: [5] © ETSI 2001.

9.4 cdma2000 QoS

Cdma2000 does not set such explicit QoS targets nor define its own classes analogous to the UMTS. In practice, it supports the same general range of applications and provides appropriate degrees of support through the use of radio link level features as outlined in Section 4.5 and the capabilities of Mobile IP.

QoS for CS core applications are based on the QoS information record in the initial signaling messages for call setup, as described in Section 4.5.1. The key factors in this are the service option, which describes the application and its bandwidth, and the selection of assured or nonassured mode over the

Table 9.4
End-User Performance Expectations—Interactive Services

Medium	Application	Degree of Symmetry	Data Rate (Kbps)	Key Performance Parameters and Target Values		
				One-Way Delay (sec)	Delay Variation (ms)	Information Loss (%)
Audio	Voice messaging	Primarily one-way	4–13	< 1 for playback < 2 for record	< 1	< 3 FER
Data	Web-browsing —HTML	Primarily one-way		< 4 per page	—	0
Data	Transaction services—high priority (e.g., e-commerce, ATM)	Two-way		< 4	—	0
Data	E-mail (server access)	Primarily one-way		< 4	—	0

Source: [5] © ETSI 2001.

radio link. At the time of the writing of this book, the main options were simply speech at rate sets 1 and 2 (see Section 10.1.1).

PS QoS is based on IP DiffServ classes [6]. These may be requested by the MS via the QoS information record, or based on its user profile 3GPP2 differentiated services class options registered with its home RADIUS server, and included in the RADIUS access accept returned by the RADIUS server to the home agent. In the case of Mobile IP, the service class has to be copied by the home agent from the IP packets to the differentiated services field of the mobile tunnel [7], while the PSDN does this for the reverse tunnel [8] (if enabled) to the home agent based on the user profile. The PDSN sends PPP packets over the R-P interface encapsulated in generic routing encapsulation (GRE) with the DiffServ class set, and usually with the aid of van Jacobsen TCP/IP header compression for efficiency; the later robust header compression for error-prone networks is not supported in early releases. Call setup is started with an IS-2000 origination message from the MS that triggers the PCF to send a registration request to the PDSN via the A11 interface (see

Table 9.5
End-User Performance Expectations—Streaming Services

Medium	Application	Degree of Symmetry	Data Rate (Kbps)	Key Performance Parameters and Target Values		
				One-Way Delay (sec)	Delay Variation (ms)	Information Loss (%)
Audio	High-quality streaming audio	Primarily one-way	32–128	< 10	< 1	< 1 FER
Video	One-way	One-way	32–384	< 10		< 1 FER
Data	Bulk data transfer and retrieval	Primarily one-way		< 10	—	0
Data	Still image	One-way		< 10	—	0
Data	Telemetry— monitoring	One-way	< 28.8	< 10	—	0

Source: [5] © ETSI 2001.

Section 8.3), then set up the A10 connection after receiving the RADIUS access accept. The PCF assigns a mobile PCF session identifier (PSI) and puts it in the GRE headers for encapsulation of the PPP packets to identify them over the R-P link. GRE headers, whose structure is shown in Figure 9.3, are stripped at either end of this link.

The significance of these fields is as follows:

- C (checksum present) is set to 0 since absent for this application;
- R (routing present) is set to 0 since absent here;
- K (key present) is set to 1 since key is present;
- S (sequence number present) is set to either 0 or 1 as apt here;

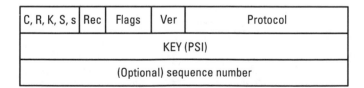

Figure 9.3 GRE header.

- s (strict routing present) is set to 0 since absent;
- Rec (recursion control) is set to 0s here since not used;
- Flags is set to 0s since not used;
- Ver (version number) is set to 0s here;
- Protocol type—this is Hex 880B here to indicate PPP, or 8881 for unstructured byte stream;
- KEY is set to the PSI for the PCF-to-PSDN connection;
- Sequence number is used for link layer protocols that require packets to be delivered in sequence (e.g., PPP). Where used, it is set to 0 for the first packet of the session, then incremented by one for each subsequent packet.

The DSCP is put in the first 6 bits of the IPv4 TOS field (or the IPv6 traffic class if v6). Most networks will map DSCPs into a limited set of classes set in the RADIUS server, so a minimal implementation might map UMTS classes as tabulated in Table 9.6, where the first digit of the AF class indicates forwarding priority and the second indicates the packet drop precedence.

Policing of traffic levels for each UMTS class might mean that within an agreed bucket level the DS class might change from AFx3 for a partially filled bucket to AFx2 as it became nearly full and to AFx1 for anything over the limit. These suggested values treat the dropping of streaming frames as less critical than those for interactive or background traffic.

The actual QoS that can be obtained depends on detailed traffic engineering for both the radio network and all fixed networks involved in a call. This is most easily controlled if the core IP network makes use of MPLS (see Section 7.4.13).

Table 9.6
DiffServ Class Allocations

UMTS Class	DiffServ Class
Conversational	EF
Streaming	AF12
Interactive	AF23
Background	AF33

The few QoS targets that are set in Release 0 for audio and video streaming are now described [9, 10].

Audio and Video Requirements

Audio codecs can operate at several rates: low bit rates (near 8 Kbps), FM-stereo quality bit rates (near 64 Kbps), and high-fidelity bit rates (near 128 Kbps).

Synchronization   For transmission of combined audio and video streams, the intermedia skew should be kept below 20 ms.

Minimum Bandwidth   The service shall be able to work at the bandwidth of 28.8 Kbps or higher for streams with video and audio contents, and bandwidth of 9.6 Kbps or higher for audio-only streams.

Play-Out Delay   The video streaming service shall be able to provide service of reasonable end-to-end delay to accommodate data transfer from the source to the mobile terminal, and shall support buffering at the terminal to accommodate transmission path degradations to a specific level. The recommended maximum play-out delay is 30 seconds.

Delay Jitter   The system shall be able to operate under delay jitter of three times the RLP retransmission time in the network with retransmission activated.

Error Rate   The service shall operate over channels with end-to-end BER of the order of 10^{-3} (for circuit-switched network services) and FER in the order of 10^{-2} (for packet-switched network services).

For videoconferencing, the targets are similar, except the play-out delay has to be much less so that end-to-end delay does not exceed 400 ms. The degree of jitter that must be compensated is up to 200 ms. The target for intermedia skew covers lip synchronization as an important case, and at 20 ms should not be noticed, unlike the upper limit for videophones (as opposed to streaming) in UMTS.

Throughput must range from 32 Kbps upwards, including the specific rates of 384 and 128 Kbps for PS and CS services, respectively.

9.5   End-to-End QoS

Many applications will involve hosts on external networks, which means that the QoS will depend on these as well as the UTRAN or cdma2000 network. The key QoS parameters to discuss are the delay, jitter, throughput, and error rate.

9.5.1   Delay

The end-to-end delay is a sum of a minimum fixed part for a given route and a variable contribution that depends on the network load. The fixed part is the sum of contributions from each processor on the path, serialization on each link, and a propagation delay due to the finite signal velocity over each link.

$$\text{Minimum delay} = \Sigma \text{ processor time} + \Sigma \text{ serialization time} + \Sigma \text{ propagation delay}$$

Processors include the mobile itself, codecs, the BTS and RNC, gateways, external routers, switches or multiplexers, and hosts. Delay is most critical for the conversational class, and almost all of these make use of codecs that are responsible for a major portion of this. Many speech codecs introduce delays of the order of 20 ms for sampling plus a few milliseconds for processing (see Chapter 10) per end, while some, such as MP3, may be as long as 150 ms. In some cases, particularly speech, transcoding will be required between different source and destination algorithms. Potential examples of this include going from AMR on the UTRAN to EVRC on a cdma2000 network, or to G.723.1 for a videoconference over a remote CS network, or to G.729 on an IP phone. In some remote networks, tandeming may also occur, where the call has to be decompressed for an intermediate call routing apparatus to read the signaling information before recompression and onwards transmission. Both transcoding and tandeming introduce delays comparable to those of the codecs (and also insert distortion), so UMTS has techniques for avoiding them [11, 12]. Routers, switches and multiplexers introduce comparatively short delays of the order of a few milliseconds. Serialization time is given by the bit size of a burst or packet divided by the bit rate of the transmission link—for example, 24 bytes of speech at 9.6 Kbps takes 20 ms, but only about 1 ms at 1.544 Mbps. One complication to this is that on a radio link, several bursts may be interleaved to reduce errors, but they cannot be forwarded until this is decoded, so the effective serialization time is that of the interleaved set instead of a single burst (e.g., 40 ms for two interleaved 24-byte bursts at 9.6 Kbps). In UMTS this interval is the TTI, which can take values of 10, 20, 40, or 80 ms; while for cdma2000 the analogous figures are 20, 40, or 80 (with 5 for urgent short signals). Speech in UMTS uses a TTI of 20 ms, allowing interleaving of two radio frames, so it does not offer any processing time advantage over cdma2000 for this type of traffic, although the 10-ms radio frame can offer shorter queuing delays (see below) for interactive traffic on the shared channels.

Propagation delay depends on link length and type according to

$$\text{Propagation delay} = \text{link length/signal velocity} \qquad (9.1)$$

where the signal velocity is that of light for microwaves (i.e., about 3×10^5 km/s in air and about 2×10^5 km/s for transmission down most types of cable). Consequently, a local call only entails the order of 1-ms propagation delay, while an intercontinental or transcontinental surface call has about 50 ms and a satellite hop via a geostationary orbit introduces a delay of 270 to 300 ms.

The minimum delay can thus vary from about 50 ms for a speech call between mobiles on the same network using similar codecs up to about a second on an international call with complex codecs.

The variable delays consist of queuing delays at each node or link on the route between source and destination. For a CS call these are negligible compared to the fixed delays; they consist of a small amount of jitter on the SONET/SDH networks that may be transited, a possible ATM AAL2 multiplexing delay, plus a small amount of queuing on the radio links if a dedicated traffic channel is not used. Speech uses a dedicated channel, but another CS application, fax, usually uses a shared channel, and so might be subject to the order of 10 ms of extra delay. The AAL2 multiplexing delay depends on mixing several distinct real-time traffic streams into a single AAL2 cell and is subject to a configurable maximum value determined by the CU Timer (see Section 7.3 and [ITU-T recommendation I.363.2]). There is no standard value for this timer, so the delay is implementation dependent and potentially about 1 to 20 ms, and at high loads the cell will be filled and transmitted before its expiry.

Rough estimates of the queuing delays can be quickly obtained by the use of classical analytic waiting time formulas, such as $M/M/1$, $M/D/1$, and $M/G/1$ [13]. In these queuing models, M indicates a Markovian or exponential distribution, D is a deterministic (i.e., fixed) value, and G is a general one; the 1 indicates the number of servers for the queue. The first letter describes the arrival rate characteristics, and the second covers the distribution of packet sizes. If $E(A)$ denotes the average arrival rate of packets, the Markovian probability for the arrival of K packets within a time T is given by the Poisson distribution $P(K)$ where

$$P(K) = \{E(A)\,T\}^K \, \mathrm{Exp}\{[-\,E(A)\,T]/K!\} \qquad (9.2)$$

The servicing time t for the packet is also exponentially distributed for $M/M/1$, so the probability $P(t)$ of a time t if the average $E(S)$ is given by

$$P(t) = \mathrm{Exp}\{-\,t/E(S)\} \qquad (9.3)$$

In the absence of prioritization the simple analytic results are summarized below. If U is the fractional utilization of the server ($0 \le U \le 1$), then the average waiting time $E(W)$ is given by

$$E(W) = U\{E(S)/(1 - U)]\} \qquad (9.4)$$

The variance of the waiting time $V(W)$ is then given by

$$V(W) = \{E(S)\}^2[(2U - U^2)/(1 - U)^2] \qquad (9.5)$$

where the general definition of variance is $V(W) = E(W^2) - [E(W)]^2$.

The definition of queuing delay used in UMTS [1] is the 95th percentile of the delay, and this can be obtained from the variance $V(W)$ if additional assumptions are made about the distribution of waiting times. If a gamma distribution is assumed, then the 95th percentile is twice the mean square root of the variance, assuming a normal distribution gives a multiple of only 1.6.

The case of $M/D/1$ is simpler and gives expectancies $E(W)$ and $V(W)$ that are lower than those for $M/M/1$ as shown below:

$$E(W) = U\{S/(2[1 - U]\} \qquad (9.6)$$

$$V(W) = S^2\{4U - U^2\}/[12(1 - U)^2] \qquad (9.7)$$

where the expectancy of the serving time $E(S)$ is just the fixed service time S.

$M/M/1$ tends to be the better for general packet networks, although slightly pessimistic due to the allowance of queuing behind nonexistent infinite packets, while $M/D/1$ is more applicable to fixed radio bursts on the shared channels and to ATM VBR services.

The $M/G/1$ model is potentially better than either if the actual size/service time distribution is known, but it has much more complex results.

$$E(W) = E(S)\{U[1 + \{V(S)/E(S)^2\}]/(2(1 - U))\} \qquad (9.8)$$

The variance of waiting time $V(W)$, is given by

$$V(W) = \{U/[12E(S)(1 - U)^2]\}\{4[1 - U]E(S^3) + 3U[E(S^2)]^2\} \qquad (9.9)$$

The most accurate results, especially at high utilizations, are obtained by network simulation that takes account of the flow control and forwarding algorithms specific to the protocols and proprietary equipment used.

For the radio link between mobile and base station, the rough queuing time for the shared channels is given by the $M/D/1$ model with a service time of 10 ms for UMTS, or 20 ms for cdma2000. This is exhibited as a MAC scheduling delay that depends on whether MAC-d or MAC-c/sh is involved. Real-time traffic that uses dedicated channels via MAC-d (e.g., speech) does not experience these delays—the small scheduling delay is absorbed by the codec processing time—but other traffic experiences delays. The rough average $M/D/1$ queuing delays and 95th percentile delays are shown in Table 9.7 below for loads of 50% and 80%.

One can estimate 95th percentiles roughly by multiplying the square root of the variance by a factor characteristic of an appropriate probability distribution. Two common choices of distribution are normal and gamma, with respective factors 1.6 and 2.0. Of these, the gamma distribution is

Table 9.7
Approximate Waiting Times

Technology	Load (%)	Service (ms)	Average Wait (ms)	Variance	95th Percentile (ms)
UMTS	50	10	5	44	13.2
UMTS	80	10	20	533	46.1
cdma2000	50	20	10	175	26.4
cdma2000	80	20	40	2,123	92.0

preferred in Table 9.7 on account of the likely asymmetric skew of the delays, while the difference between the two factors gives a rough indication of the errors inherent in this technique. On the radio-link prioritization is normally applied, and with preemptive priorities the average wait for the top level is given by the same formulae, but with the utilization factor for the top level only. If this prioritization were applied to the interactive class as a whole with a class utilization of 10%, then at a total of 50% or 80% load including background class, the average wait for interactive would only be 1.1 ms for UMTS or 2.2 ms for cdma2000. Accurate results require detailed simulation of the exact behavior of the system.

On a PS network the variable delays are potentially much larger if there are any slow links or large packets, although the trunks of an IP core network should be too fast to cause a problem. The serialization time for various packet sizes and line speeds is given in Table 9.8.

In Table 9.8 (1) refers to a packet of 40 bytes (e.g., compressed voice), (2) is a 256-byte packet, and (3) is a maximum IPv4 MTU of 1,500 bytes. The waiting time is based on an $M/M/1$ model with average packet size of 256 bytes at a typical busy hour design loading of 50%. This shows that serious delays are only likely from the access lines to devices in small offices on remote PS networks. With IPv6 (see Section 7.4.2) the minimum MTU is 1,280 bytes, so link-level fragmentation may be needed on these slow links.

Calls from one mobile to another, even if it is on a different PLMN, will be free from these low-speed access lines, and should only experience PS delays no worse than those for Wait (2) on T1/E1 lines. Use of DiffServ class EF for real-time traffic means that their delays should be significantly less than the average thanks to prioritized PHB within any IP networks, while

Table 9.8
Serialization Times

Line Type	Line Speed [Kbps]	Time (1) [ms]	Time (2) [ms]	Time (3) [ms]	Wait (2) [M/M/1 @50%]
T3	45,000	0.007	0.04	0.26	0.04
E3	34,000	0.009	0.06	0.35	0.06
E1	2,048	0.15	1.0	5.8	1.0
T1	1,544	0.20	1.3	7.7	1.3
Nx64	512	0.6	5.2	23.4	5.2
Nx64	256	1.25	8.0	46.9	8.0
DS0	64	5.0	32.0	187.0	32.0

background traffic with a lower AF class will experience longer waits on any such network and will face the shared radio channel delays as well.

Traffic that requires a low error rate and uses the RLC AM will also experience some retransmission delays. The average value of this delay is the fraction of frames retransmitted times the sum of transmission time and retransmission timer.

The 95th percentile transfer delay used in UMTS QoS profiles is the sum of the fixed delay components plus the overall 95th percentile queuing/retransmission delay.

9.5.2 Delay Variation

Delay variation is cumulative across the network as indicated in (9.10) below, but it can be compensated at the application level by using a play-out buffer and synchronization fields in the protocol. The external networks are expected to use either ATM with AAL5 or the Internet multimedia core with IP. In the case of ATM, the CDV is one of the parameters that is specified with QoS categories VBR.1 and VBR.2, although it is only optional for the latter. In the case of IP, there is no negotiable delay variation parameter, even with IntServ and controlled delay, so the network has to be engineered to provide acceptable parameters. UMTS does not attempt to specify the delay variation directly in its QoS signaling parameters, but does specify a 95th percentile delay figure rather than the average, so that some account is taken of fluctuations. The 95th percentile delay figures for separate network components are not directly additive, but are based on the square root of the sum of variances for the components

$$T_{\text{total}} = \text{SQRT}(\Sigma_j V_j) \qquad (9.10)$$

where summation runs over all network components, and the formula assumes that the delays on the component networks are mutually independent. The 95th percentile is then given by multiplication of the root total variance by a factor of 1.6 or 2, depending on the probability assumptions made, and adding it to the fixed delay. Because of the squaring of service time implicit in variances in (9.10), the total variance is dominated by the largest terms. Again, simulation of the actual network will give much more reliable results than the rough analytic formulae and probability distribution assumptions.

Play-out delay can occur at intermediate points in the network as well as in the end codecs; for example, if an RTP session is terminated at a gateway, then the traffic is likely to be resynchronized using the RTP time-stamps before onwards transmission. In that case it provides a relatively constant additional delay component that reduces the overall jitter.

9.5.3 Throughput

In the absence of retransmissions, the throughput of a sequence of network sections is that of its slowest component, the bottleneck. If a protocol is used that entails end-to-end retransmission for error correction, then the throughput is reduced due to gaps in transmission while waiting for acknowledgments plus the retransmission overhead. If the window size and transmission rate permit pipelining, then the reduction in throughput is simply the retransmission overhead. The window size WS needed for pipelining to be achieved is given by

$$WS = RTT \times bandwidth \qquad (9.11)$$

where RTT is the round-trip time, and bandwidth is that of the slowest link in the case of end-to-end error retransmissions. This bandwidth will normally correspond to the guaranteed bit rate, where specified, for UMTS. The same formula also holds for individual links that use their own link acknowledgments and retransmissions (e.g., RLC AM in UMTS).

UMTS requests both a peak throughput and a guaranteed throughput that can be negotiated with each component network, provided that the latter uses protocols such as RSVP or ATM. For CS services the guaranteed throughput should be easily achieved, but in the case of IP DiffServ for the PS domain, negotiation is not possible and the best that can be done is to have SLAs with the external networks such that a GGSN should be provided with a certain guaranteed throughput for each of the DSCPs that are in use.

9.5.4 Errors

Errors are cumulative over the network components but can be reduced if an end-to-end error correcting protocol (e.g., TCP/IP) is being used. This is likely to be the case for interactive and background classes where data integrity is important, but delay is not. The QoS profiles (see above) cover residual BERs ranging from 5×10^{-2} to 6×10^{-8}, the BER for a fiber optic circuit is of the order of 1×10^{-10}, and a wide area digital cable should be no worse than 1×10^{-6}, so the radio link is usually the only significant factor for most traffic types.

The BER for a radio frame is normally in the range 10^{-1} to 10^{-3}, so many applications will require retransmissions to achieve their targets. Real-time streaming applications designed for wireless networks will probably need one radio retransmission on average if they are to reach their minimum standard, while those that are not so designed will need two or three. Similarly, most interactive and background applications need a low error rate, and hence will probably require three or four radio retransmissions. The mechanisms for error control at radio link level were described in Chapter 3 for WCDMA and Chapter 4 for cdma2000.

Many applications contain distinct traffic streams with separate levels of fault tolerance [e.g., AMR speech (see Section 10.1) and compressed video (see Section 10.2)], and this need is met by use of distinct transport channels over the radio links and by different DiffServ classes over IP and potentially different cell loss priority over ATM.

9.5.5 Call-Failure Probability

This feature is not part of the UMTS QoS profile but is vital to user satisfaction. It is determined by network design (see Chapter 6) and by congestion control, which takes account of potential overload situations. Mobile networks must handle large increases in network traffic as a result of frustrated transport users sending messages to say that they are stuck and will be late etcetera, and this feature could be exacerbated on 3G networks by such users attempting to access multimedia services to pass the time in areas where the demands for these are normally low. Admission control in such areas needs to be able to reserve bandwidth for the low bandwidth services at the expense of multimedia in such situations. A further risk on 3G networks is denial of service attacks based on flooding the network with video packets. The additional measures for preventing this include the use of encryption within the multimedia applications and secure operating systems on the servers.

References

[1]   ETSI TS 123 107, "QoS Concept and Architecture," V.4.3.0 (2001), http://www.etsi.org.

[2]   Third Generation Partnership Project TS23.207, "End-End Quality of Service," http://www.3gpp.org.

[3]   ETSI TS 124 008, "UMTS Radio Interface Layer 3 Core Network Protocols – Stage 3," V.4.4.0 (2001), http://www.etsi.org.

[4]   ETSI TS 122 060, "General packet Radio Service (GPRS)—Service Description; Stage 1," V.4.2.0 (2001), http://www.etsi.org.

[5]   ETSI TS 122 105, "Services and Service Capabilities," V.4.2.0 (2001), http://www.etsi.org.

[6]   3GPP2 P.S0001-A, "Wireless IP Network Standard," http://www.3gpp2.org.

[7]   IETF RFC2002, "IP Mobility Support," 1996.

[8]   IETF RFC2344, "Reverse Tunneling for Mobile IP," 1997.

[9]   3GPP2 S.R0022, "Video Conferencing Services—Stage 1," http://www.3gpp2.org.

[10]   3GPP2 S.R0021, "Video Streaming Services—Stage 1," http://www.3gpp2.org.

[11]   ETSI TS 123 153, "Out of Band Transcoder Control," V.4.2.0 (2001), http://www.etsi.org.

[12]   ETSI TS 128 062, "Tandem Free Operation of Speech Codecs—Service Description," V.4.1.1 (2001), http://www.etsi.org.

[13]   Medhi, J., *Stochastic Models in Queuing Theory*, San Diego, CA: Academic Press, 1991.

10

Specific Applications

10.1 Audio Applications

10.1.1 CS Voice

Pulse Code Modulation

Originally all voice traffic was carried over CS networks using a mixture of analog and digital links. Analog voice is converted to digital by means of pulse code modulation (PCM) as defined by the standard ITU G.711. Voice in the frequency range 300 to 3,400 kHz is converted to 8-bit samples at a rate of 8,000 samples per second, so generating a 64-Kbps digital data stream. SNR increases with the number of bits per sample, and 8 bits was originally the minimum found to give subjectively good quality voice. Two different algorithms are defined for this quantization, A-law and μ-law. The former is used in Europe, and the latter is used in North America and Japan, with responsibility for interconversion resting with the μ-law end for inter-communication. Some distortion is introduced in each conversion cycle of analog/digital/analog, and this is defined to be one quantization distortion unit (QDU) for G.711, with an overall limit of 14 QDU—the maximum acceptable end-to-end total on carrier quality international calls. The total is roughly additive per quantization cycle where tandeming occurs, so if end-to-end digital routing is employed to avoid this, there is considerable scope for using lower bandwidth algorithms with higher levels of distortion. Reduction of bandwidth from 64 to 32 Kbps by use of G.721 adaptive pulse code modulation (ADPCM) has about 3.5 QDU per quantization cycle and is widely used in CS networks, giving quality almost as good as PCM. This compares the actual signal with that predicted from the previous sample on the basis of normal speech characteristics and quantizes the differences for transmission (see [1] for general voice compression principles).

Compression Algorithms

On mobile phone networks cochannel interference prevents high signal-to-noise ratios from being used, and low-bandwidth high-compression algorithms are employed on 2G networks. For example, GSM uses a residual pulse excited linear predictive code that generates about 13 Kbps total traffic with distortion of 7 or 8 QDU, which is quite adequate for a single quantization cycle but potentially poor if additional conversion has to be used. The distortion, as measured by QDUs, is not additive for these algorithms unlike ADPCM, and quality tends to be measured in terms of a subjective mean opinion score (MOS) for a single end-to-end quantization cycle. The MOS is a value in the range of 1 to 5 based on the views of a random sample of people according to the criteria in Table 10.1.

Different samples of people will come up with differing values, so a more objective measure is needed. ITU P.862 is such a recommendation for standard test operations. A very large number of voice compression algorithms are in use, and the rough MOS and bandwidth used are indicated in Table 10.2.

In Table 10.2, the asterisk following the mobile phone compression algorithms indicates the best MOS possible, which is achieved at an SNR of at least 25 dB. As the SNR drops, the MOS falls roughly linearly to the order of 1 at 10 dB. GSM Enhanced Full Rate (EFR) is an improved version of GSM RPE-LP whose half-rate version established a poor reputation. The distinction between G.729 and G.729a is that the former is a more

Table 10.1
MOS Quality Criteria

MOS Scale	Subjective Quality
5	Excellent
4	Good
3	Fair
2	Poor
1	Unacceptable

Table 10.2

Voice Compression and Quality

Algorithm	Bandwidth (Kbps)	MOS
G.711 PCM	64.0	4.2
G.721/726 ADPCM	32.0	4.0
G.728 LD-CELP	16.0	3.6
GSM RPE-LP (FR)	13.0	4.0*
GSM EFR	12.2	4.1*
G.729 CS-ACELP	8.0	3.9
G.729a CS-ACELP	8.0	3.7
G.723.1 MP-MLQ	6.3	3.9
G.723.1 ACELP	5.3	3.7

Note: Low delay-code excited linear prediction (LD-CELP); conjugate structure algebraic CELP (CS-ACELP); multipulse multilevel quantization (MP-MLQ); full-rate/half-rate (FR/HR).

exhaustive and processor-intensive algorithm than the latter. In G.723.1, the audio algorithm that is recommended for multimedia in ITU H.324 (the low-bandwidth option) is intended for use when network congestion is detected, and codecs should be able to support both at the same time in opposite directions.

Quality falls if the voice has to be either tandemed or transcoded; for example, tandeming of G729 leads to a drop in MOS to about 3.3 and transcoding is liable to be even worse. A very large number of voice compression algorithms have been used on 2G mobile networks, and many are listed in Table 10.3.

This multitude of options means that the choice of codec must either be negotiated at call setup, or else transcoding provided to allow communication at a degraded quality. Where transcoding is necessary, it is usually performed at the MSC using G.711 as a common intermediate standard between the two networks. On GSM networks, EFR is a replacement for the original RPE-LP algorithm, which often gave poor quality, particularly at half-rate. The different rates for QCELP on IS-95 CDMA networks allow for discontinuities in speech rather than changes in reception (sending frames of 266, 124, 54, or 20 bits, respectively). The later EVRC algorithm uses the same overall principles, but with a higher bandwidth (and quality)

Table 10.3
2G Voice Compression Algorithms

Algorithm	Network(s)	Bandwidth (Kbps)
VSELP (HR)	GSM/DCS1800/PCS1900	6.5
RPE-LP (FR)	GSM/DCS1800/PCS1900	13.0
ACELP (EFR)	GSM/DCS1800/PCS1900	12.2
VSELP (FR)	IS-54/IS-136	7.95
ACELP	IS-54/IS-136	7.4
ACELP (EFR)	IS-54/IS-136	12.2
VSELP	PDC	6.7
PSI-CELP (HR)	PDC	3.45
QCELP	IS-95	8, 4, 2, 1
RCELP	IS-95	Variable (EVRC)

RCELP codec for the CDMA rate set 2 hierarchies of speeds based on 14.4 Kbps instead of the 9.6 Kbps of rate set 1 used with QCELP. There is a standard way of carrying QCELP in RTP (see Section 7.4.7 and [2]).

3G UMTS aims to avoid this problem by largely standardizing on a single codec, UMTS adaptive modulation rate (AMR) [3], which is derived from ACELP (EFR) and an earlier GSM AMR codec. This has a maximum bit rate of 12.2 Kbps but operates at lower bit rates of 10.2, 7.95, 7.4, 6.7, 5.9, 5.15, and 4.75 Kbps when the received signal quality is high enough to provide adequate quality of service at the higher degrees of compression. These voice calls normally use a full-rate CS traffic channel of 22.8 Kbps, so that the difference of 10.6 Kbps and the EFR rate of 12.2 Kbps consists of the redundancy information introduced in the channel coding. When signal quality drops, a higher degree of compression is used to allow for a greater degree of redundancy and error correction. In general, the intermediate AMR rates do not match any other codecs, but the 7.4-Kbps rate matches the IS-641 EFR algorithm for speech on American IS-136 TDMA networks.

AMR also distinguishes between the importance levels of different aspects of speech through the use of three classes with different target residual bit error rates. This codec is likely to be added to the IS-95 list for cdma2000 also. When a 3G mobile phone contacts a 2G network, it must negotiate the codec to be used from a defined list [4], but UMTS AMR is automatically used when connecting to other 3G UMTS networks. Transcoder free operation (TrFO) is achieved by out-of-band signaling between the source and destination MSCs (or associated media gateway controllers) [5]. The initiating UE sends a list of supported codecs to its MSC, which passes this on to the MSC for the destination UE. The called UE passes its own list of supported codecs back to the destination MSC, which then selects the codec to be used.

Even with a single codec, transcoding is liable to be used on calls between different PLMN if the internetwork link does not support compressed voice. The basic mode of operation is for the MSC in one PLMN to convert the compressed voice to G.711 PCM A-law or μ-law for transmission to the MSC of the second PLMN, which then recompresses the voice for transmission to the second mobile. This causes both delay and loss of quality, so both GSM and UMTS have specific mechanisms for both TrFO and tandem free operation (TFO) [6]. Unlike TrFO, TFO uses a special protocol that runs between the transcoders in the MSCs, or in the case of 3G networks the transcoders in the media gateways for the MSC servers. The initial version of TFO supports the GSM_FR, GSM_HR, GSM_EFR, and four AMR speech codec types (FR_AMR, HR_AMR, UMTS_AMR, UMTS_AMR_2). At the start of the call some PCM samples have to be exchanged between the transcoders and any mismatch between codecs resolved before TFO can be implemented. Once TFO is in use, only 8k or 16 Kbps (depending on the codec) is used for the compressed voice on the 64k PCM channel. TFO has many detailed features for specifying frame structures, setup, handovers, etc., as well as specialized circumstances such as rate control for the AMR codecs.

The preferred codec in UMTS for both TrFO and TFO is UMTS_AMR_2, which is a superset of UMTS_AMR and EFR, thereby enabling interoperation with GSM mobiles. EFR is used on the UL, and UMTS_AMR on the DL in this case. If the codec lists are restricted to UMTS only, then UMTS_AMR is used in both directions.

Echo

A common feature of the high-compression algorithms is that most make use of silence suppression. In a normal full-duplex conversation about 60% consists of silences due to pauses and one person speaking at a time, so encoders try to eliminate these, while the decoders add a small amount of background comfort noise into the silent gaps. Avoidance of clipping the start of talk-spurts entails buffering of several samples while they are analyzed, so that any containing speech can be transmitted once it has been detected. This, together with the general processing time for compression or decompression, causes a delay that can lead to echo, which must be eliminated. The echo arises from any location in the overall voice circuit at which there is a sharp change in impedance and consequent partial signal reflection—a traditional example of this being an analog 2-wire/4-wire hybrid in PSTN—and is detectable if the delay exceeds about 30 ms. As a result echo cancellers are used that estimate the size and delay of the echo based on the locally transmitted sound and circuit characteristics, and subtract it from the received signal. Echo cancellation is normally applied to each talkspurt separately, and there is a brief period at the start of each during which the echo canceller converges and a declining echo is likely to be heard. Standard recommendations for echo canceller performance are ITU G.164 and G.165 for analog circuits and G.168 for digital. G.168 recommends that convergence should be at least 16-dB echo return loss (ERL) within 1 second, and many current echo cancellers easily exceed the specification; the total ERL between a speaker's mouth and ear that must be compensated (G.131) is likely to be 46 to 54 dB. A digital cellular handset behaves like a 4-wire circuit as regards echo, but also suffers from some nonlinear delays associated with low bit rate compression and also from acoustic cross talk between the loudspeaker and microphone within the handset. On analog circuits a 2,100-Hz tone is used to detect modem or fax traffic and disable the cancellation. Echo cancellation is confused when both parties speak at the same time, and cancellers have to have the ability to detect and allow for this doubletalk. A further complication is that the impedance characteristics of circuits differ throughout the world, so settings have to be tweaked to allow for this too.

The consequence of this is that a mobile network normally uses an echo canceller specific to the equipment types and radio link delays in the mobile switching center (or gateway) that interfaces the PSTN. UMTS recommends a G.168 compliant echo canceller at this point to handle the echo from 2-wire to 4-wire conversion [7]. Echo associated with cross talk tends to be compensated in the transcoder (rather than in the echo canceller) along with the other nonlinear delays resulting from the compression algorithm. The transcoder required to go from the low bit rate voice to G.711 PCM is also situated in the MSC.

Several of the compression algorithms for mobile networks produce sets of output bits with distinct degrees of importance, which are then treated differently over the radio links. An early example of this is the original RPE-LP GSM FR where there are three classes of bit. The algorithm produces bursts of 260 bits every 20 ms (hence 13 Kbps), of which 50 bits are designated Class 1a, 132 as Class 1b, and 78 as Class 2. The Class 1a bits are supplemented by 3 parity bits, then after some interleaving they and the

Class 1b bits are encoded with a rate 1/2 convolutional code, and then the Class 2 bits are added unencoded to produce the frame for transmission. If the Class 1a bits are found to be corrupted, the whole frame is dropped, but errors in the others are acceptable. A similar procedure is adopted in the improved EFR and is taken a degree further in UMTS_AMR. Here the sizes of the bursts and the number of bits in the three classes depend on which of the possible source compression rates is being used. High source rates are used with high convolutional code rates (e.g., 1/2) when network quality is good; as network quality drops, however, the source rate is reduced and the convolutional code rate is lowered (e.g., to 1/3) to give more error correction for the optimum practical result.

UMTS_AMR

Since UMTS_AMR is the main codec for UMTS, the way in which it uses the radio bearers and the acceptable levels of error have been thoroughly investigated. Like the earlier GSM algorithms, this has three RAB subflows corresponding to the three classes of bits, each of which leads to a dedicated transport channel.

AMR operates at a rate of 50 20-ms frames per second, but the number of bits per frame depends on which of the eight possible rates is in use. These frame sizes are mandatory, but the subdivision into the three classes, A, B and C, is only optional with suggested values as in Table 10.4 [8]

A call will not necessarily permit use of all of these rates, and an active subset of those available for use is specified. For each member of the active subset a RAB format combination indicator (RFCI) is defined; together

Table 10.4
AMR Frame Sizes

Rate (Kbps)	Frame Size	A Bits	B Bits	C Bits
12.2	244	81	103	60
10.2	204	65	99	40
7.95	159	75	84	0
7.4	148	61	87	0
6.7	134	58	76	0
5.9	118	55	63	0
5.15	103	49	54	0
4.75	95	42	53	0

these form the RAB format combination set (RFCS). The permitted sub-frame sizes for each quality class also define the block sizes in the TFS for the mapping of transport channels to the physical channels of the radio link (see Section 3.3.4). For example, if the active subset consisted of only 12.2- and 7.4-Kbps rates, plus 0 Kbps (for silences) and a silence descriptor rate for describing the characteristics (e.g., insert background noise) of silences, the RFCS would be as shown in Table 10.5; if all possible rates were permitted, the TFS for each channel would include all the entries in each of columns A, B, and C in Table 10.4, plus entries associated with silences.

The figures in the columns in Table 10.5 provide the SDU sizes for the SDU format indicators for each subflow in the RAB parameters (see Section 9.2). The RAB parameter IE also needs the bit rates (maximum 12.2 Kbps, guaranteed 7.4 in this case), error rates, and delivery behavior for erroneous frames. The values for these fields are left to the operator implementation, but with plausible values as indicated in Table 10.6.

The SDU error ratio is only applicable to subflow 1 as it is the only one whose frames, sent as I_u PDU type 0, are provided with a CRC frame quality indication set by the level 1, the others being PDU type 1 (see Section 8.1.3.).

Table 10.5

Example of an RFCS for AMR

RFCI	Subflow 1	Subflow 2	Subflow 3	Total	Source Rate
0	81	103	60	244	1
1	61	87	0	148	2
2	0	0	0	0	3
3	39	0	0	39	4

Table 10.6

Possible Error Parameters for AMR Subflows

Subflow	Residual BER	SDU Error Ratio	Error Delivery
1	1×10^{-6}	7×10^{-3}	y
2	1×10^{-3}	—	—
3	5×10^{-3}	—	—

The AMR subflows are passed down to level 2 under RRC control where they all use RLC transparent mode, but are passed to the MAC layer for separate DCHs. In this mode RLC only performs segmentation and reassembly of the I_u PDUs so that they fit into suitable sized RLC PDUs for transmission at the correct rate. These are passed to the MAC layer where they are again handled transparently. Data is passed between the MAC layer and the physical layer via transport blocks. In the case of AMR, the block sizes for each of the three DCH are simply those for the numbers of bits for the corresponding class in Table 10.5. The transport format also indicates the TTI of 20 ms for each, the type (here convolutional) and rate of the channel coding to use, and the CRC size to use for each. The channel coding rate depends on which AMR bit rate is used, as the purpose of the variable rates is to allow the code rate to be increased if signal reception is poor. For the class A bits at 12.2 Kbps, the code rate is 1/2, while B and C bits are unencoded, but at 7.4 Kbps the A bits use a rate 1/3 code, and B uses 1/2. Only the class A bits use a CRC, and this is an 8-bit field.

The end-to-end delay target for AMR speech is the 100-ms limit needed to facilitate the speakers from not interrupting each other.

UMTS speech normally uses the CS core network with signaling for ATM AAL2, but when a conversation is held with an IP phone then either the PS domain or the Internet multimedia core might be involved instead. If so, the call will entail use of a media gateway for transcoding, while signaling is handled by the MGCF (see Chapter 8) through SIP commands. The session is established by a SIP INVITE that details the call features through SDP elements, followed by SIP series 1xx call progress signals and a SIP OK 200, with eventual release by a SIP BYE. A typical SDP description of AMR would be

$$m = \text{audio } 49120 \text{ RTP/AVP } 97$$

$$a = \text{rtpmap: } 97 \text{ AMR/8000}$$

$$a = \text{fmtp: } 97 \text{ mode-set} = 0,2,5,7; \text{ mode-change-period} = 2;$$
$$\text{mode-change-neighbor; maxframes} = 1$$

AMR can be carried in RTP using special procedures [9], and this is indicated in the first line above, where 97 represents a dynamic RTP payload type and 49120 is the UDP port number. AMR/8000 indicates the normal sampling rate of 8,000 per second as opposed to the less common wideband

AMR [10]. The mode-set indicates which of the eight AMR rates are to be supported by the session (in this example, 4.75, 5.9, 7.95, and 12.2 Kbps). The main complication of AMR for RTP is the need to change mode (i.e., rate) from time to time for QoS reasons. The mode-change-neighbor entry means that the rate can only change from its current value to an adjacent member in the mode-set, while the change-period restricts the frequency of such changes. Maxframes is the maximum number of AMR speech frames that can be put in a single RTP packet; increasing this parameter causes an extra packetizing delay. The time-stamp of the RTP packets with one frame per packet increases by the sampling number of 160 in successive packets. There is a payload header following the RTP header that specifies the index (0–7, as in the mode-set) of the rate appropriate to the individual packet.

In general, transcoding to another algorithm, such as G.729a, will be required for a session to a VoIP user. This will entail a separate RTP session on the IP PSDN.

Wideband AMR (AMR-WB) [10] is a later addition to the UMTS codec range that doubles the sampling rate from 8,000 to 16,000 per second and uses more bandwidth than would be available on the original GSM FR and HR traffic channels. It has nine rate options, as shown in Table 10.7, and only uses class A and B bits.

AMR-WB is likely to be used (with transcoding) for communication with VoIP devices via the IM.

Table 10.7
AMR-WB Rates

Index	Rate (Kbps)	Total Bits	A Bits	B Bits
0	6.6	132	54	78
1	8.85	177	60	117
2	12.65	253	72	181
3	14.25	285	72	213
4	15.85	317	72	245
5	18.25	365	72	293
6	19.85	397	72	325
7	23.05	461	72	389
8	23.85	477	72	405

10.1.2 PS Voice

There are two main types of packetized voice in addition to ATM: VoIP and Voice over Frame Relay (VoFR), of which the former is much more important. The initial business drivers for each of these were the cost savings resulting from the use of highly compressed voice over a private or virtual private network instead of the public network—while VoIP also allows integration of speech and data in Web applications together with some cost savings where VoIP exchanges are used instead of traditional PABXs. In contrast to ATM, which was designed to provide high voice quality through the use of 48-byte cells at transmission rates of 51 Mbps or greater, VoIP and VoFR are used on links carrying much larger packets at speeds that may be as low as 64 Kbps and so are subject to large and variable delays due to queuing behind data packets. The user of a 3G mobile network may speak to a remote VoIP or VoFR user on another network, and so can be affected by the difficulties associated with these protocols.

The magnitude of these delays is shown in Table 10.8 for a range of line speeds from 64 Kbps to 45 Mbps and a range of data packet sizes for the case where a voice packet is queued immediately behind a complete data packet. The table also shows the serialization time for transmission of a typical voice packet of 24 bytes of compressed speech plus 4 bytes of compressed RTP/UDP/IP header, or 64 bytes with uncompressed header.

The MTU for data packets is at least 256 bytes for IPv4 and at least 1,280 bytes for IPv6, so fragmentation is normally required on low-speed lines (i.e., less than about 1 Mbps) for acceptable voice delays. The Frame

Table 10.8
Packet Delays

Line Speed Size	Delay (ms)							
	64	256	512	1,024	1,544	2,048	34M	45M
28	3.3	0.8	0.4	0.2	0.14	0.1	0.006	0.004
64	8.0	2.0	1.0	0.5	0.33	0.25	0.015	0.011
256	32.0	8.0	4.0	2.0	1.3	1.0	0.06	0.044
640	80.0	20.0	10.0	5.0	3.3	2.5	0.15	0.11
1,000	128.0	32.0	16.0	8.0	5.2	4.0	0.24	0.17
1,500	187.0	47.0	23.5	12.0	7.9	6.0	0.36	0.25

Relay Forum (FRF) defines a standard, FRF.12 [11], for the fragmentation procedure for frame relay, and the same technique is also adopted for VoIP, but the actual size of fragments is not specified. One approach sometimes used is to make the data fragments the same size as the voice packets (which in the example above is 28 bytes), but this degree of fragmentation puts an excessive processing load on the routers that execute this, as well as producing a big header overhead. Consequently, a more practical approach is to use a fragment size of about 80 bytes per 64-Kbps line speed, so that the characteristic queuing delay is 10 ms. The fragmentation is performed on a link-by-link basis, with the receiving router using the RTP time-stamp to eliminate most of the jitter.

Unless the voice packets are prioritized over data, they could just get stuck behind a sequence of data fragments, so a queuing mechanism must be adopted. The queues must be examined at least once every 10 ms. For an access line where it is known that the peak speech rate cannot exceed about half the bandwidth available, voice can be given preemptive priority so that it never has to wait for more than one data fragment (though it might wait for several other speech packets), but more generally some form of fair queuing must be used. Typically, this entails giving voice a high IP precedence class (e.g., 5, which is less than network management commands) or DSCP 101110 for premium service (EF PHB), and allocating guaranteed average bandwidths to other classes on the basis of their levels and expected volumes.

Compression and packetization of the voice introduce a delay of the order of 20 to 30 ms, and again echo cancellation must be provided to counter this, with the most effective location for doing so being at the source of the delay.

RTP distinguishes between packetized voice types according to the compression algorithm via the RTP payload type, as indicated in Table 10.9 for some of the more common algorithms.

Table 10.9
RTP Audio Payload Types

Algorithm	Payload Type
G.721/726 ADPCM	2
GSM	3
G.711 PCM-A Law	8
G.728 LD-CELP	15

Where an algorithm supports silence suppression (e.g., G.728), RTP sets the marker bit in the header for the start of each talkspurt. Multichannel audio with unframed samples is packetized by inserting octets in a left-to-right channel sequence with common time-stamp. Many algorithms use a payload type from RTP's dynamic range of 96 to 127 instead of a fixed value.

RTP in its original form does not support adaptive rate algorithms such as AMR, and special modifications are needed to handle this (see above for AMR and [9]).

For mobile network users, communication with a device using VoIP is a worst-case scenario. The first problem is the incompatibility of codecs, which leads to both a transcoding delay of the order of 50 ms (i.e., half the delay budget) together with a considerable loss of quality in the transcoding. The second problem is the variability of the additional delay over the shared IP network links, which introduces a stuttering effect. The third problem is the difficulty in tuning the extra echo cancellers that are required.

10.1.3 MP3 Audio

The voice compression standards described above are totally inadequate for hi-fi music, for which the most widely used method is MP3 compression. This is not an independent standard, but the most highly compressed of three options for the audio component of broadcasts in the original MPEG-2 standard [12]—MP3 stands for MPEG level 3.

The characteristics of these three levels are tabulated in Table 10.10, which shows the typical bandwidth used for a 1.5-Mbps soundtrack from a CD.

Fixed-rate audio samples are transformed into frequency bands, and components that the human ear will fail to detect are removed along with redundant information. One major feature of this is that a loud sound on one frequency will mask adjacent frequencies, so that these can be dropped.

Table 10.10
MPEG-2 Audio Levels

Level	Rate (Kbps)	Maximum Delay (ms)	Minimum Delay (ms)
MP1	192	50	19
MP2	128	100	35
MP3	64	150	50

The algorithms make use of extensive tables of these effects. The higher the degree of compression, the more processing and buffering are required; hence, these are greatest for MP3. MP3 is the most sophisticated of the three and gives higher quality than MP2, which is based on the earlier MUSICAM standard.

MPEG-2, and hence MP3, supports five compressed audio channels: left, center, right, right surround rear, and left surround rear, plus a low-frequency subwoofer channel. It also has an option for seven audio channels for language dubbing. Part of the compression for all three levels comes from combining the two left and two right channels into a master and slave, with elimination of redundancy; while MP2 and MP3 also delete stereo effects for frequencies greater than 2 kHz. The reason for the latter is the inability of the human ear to determine the direction of sounds at frequencies above 2 kHz.

MPEG-2 was originally designed to use ATM for network transport and uses 188-byte packets that fit into four AAL1 cells, but MP3 is almost always carried in IP instead using RTP encapsulation. The procedure for doing this was defined in [13], with suggestions for improvement in [14]. Audio is encapsulated differently in RTP to video, as indicated by a separate RTP payload type. At a common sampling frequency of 44.1 kHz, a single audio frame lasts about 26 ms; so a typical MP3 compressed data rate of 64 Kbps produces very roughly 200 bytes of data on average. The sizes of individual compressed frames vary because of differences in their compressibility and content, so 188-byte MP3 frame boundaries do not coincide with audio frame boundaries. The specification states that a single packet should contain an integral number of audio frames, and the frame boundary and-packet boundary must coincide. In order to control the effects of jitter, a 32-bit time-stamp is included for the audio frame based on the 90-kHz clock of the encoder. Following the RTP payload type in the 4-byte audio header, there is a reserved field, then a fragment offset that defines where the audio data starts in the packet. MP3 frames are characterized by RTP payload type 14.

MP3 (but not level 1 or 2) frames contain a back-pointer to earlier frames and require them for decoding; and the frame/packet boundary coincidence for MP3 causes a problem if an MP3 packet is lost, as is likely on error-prone networks. As a result, the later revision [14] is based on application data units, corresponding to the audio frames that contain this pointer rather than the MP3 frames. The RTP payload type for this revised encapsulation has to be different to the original value of 14, and is recommended to be from the RTP dynamic range of 96 to 127.

10.2 Video Applications

10.2.1 Video Basics

Video applications make use of compression algorithms and some error correction techniques in order to allow efficient storage or transmission across networks. The images for transmission can be either static or mobile, leading to different algorithms. The main static standards are Joint Photographic Experts Group (JPEG) for color or gray-scale pictures and Joint Bi-Level Group (JBIG) for black and white. Motion standards include ITU-T H.261 video-conferencing, H.263, and MPEG-1, -2, and -4. The early standards, such as JPEG and MPEG-1 and -2, are not likely to be implemented on mobile terminals but are included here as they may be encountered by users who plug palm-top and laptop computers into mobile adaptors. Their weaknesses are also helpful in understanding the later applications designed for the mobile environment. The list of multipurpose Internet mail extension (MIME) types [15] that are specifically recommended in UMTS [16] are AMR narrowband and wideband (see Section 10.1.1); MPEG-4 AAC audio; MPEG-4 video (see Section 10.2.6) and H.263 video (see 10.2.7) for continuous multimedia; JPEG (see 10.2.2); graphics interchange format (GIF) for bitmaps; and extensible hypertext markup language (XHTML) media types.

Apart from JBIG [17], each of these standards takes as its starting point an image that is related to that produced by a CCIR 601 broadcast-quality TV camera [18]. The image is described in terms of luminance and chrominance format (Y, U, V), where Y represents the brightness, U the color in a horizontal direction, and V the color in a vertical direction. A picture that is sampled at equal resolution in the luminance and chrominance planes is described as having format $(4, 4, 4)$, but as the human eye is more sensitive to changes in brightness than color, the chrominance resolution can be reduced relative to the luminance without noticeable degradation of the image. This leads to the two more usual formats $(4, 2, 2)$ and $(4, 2, 0)$. Format $(4, 2, 2)$ entails sampling the chrominance plane horizontally at half the luminance resolution, while $(4, 2, 0)$ has half resolution for chrominance both horizontally and vertically. The CCIR 601 image has format $(4, 2, 2)$ with 720 pixels per line, 243 lines, and 60 fields per second, while the compression algorithms work on the half resolution $(4, 2, 0)$ format. This basic source input format (SIF) unfortunately varies according to whether the source is for Europe (360 pixels per line, 288 lines per frame, and 25 frames per second), where two fields are also combined into a single frame, or North America and the Far East (360, 240, 30); as a result, a common intermediate format

(CIF) is used for compression. CIF is of format (4, 2, 0) with 352 pixels per line, 288 lines per frame, and nominally 30 frames per second (actually 29.97). The (4, 2, 0) format means that there are 288 lines per frame for luminance in CIF, but only 144 for chrominance. The low transmission rates of networks, particularly PSTN, N-ISDN and mobiles, means that lower resolution images, such as QCIF (half the number of pixels per line and half the number of lines per frame of CIF), are the main formats relevant to 3G networks.

The compression algorithms are also optimized for the human eye and are probably not ideal for use by automatic computer analysis applications. All but MPEG-4 are based on block compression algorithms in which pixels are grouped into blocks, then macroblocks and groups of macroblocks. The pixel information is put into luminance/chrominance format, and this data for the blocks is then transformed by a discrete cosine transformation (DCT) analogous to a Fourier transform. This expresses the data as a constant plus a series of coefficients multiplying a set of cosines of different frequencies, usually with 64 elements. Gradual changes in luminance or chrominance produce mainly low-frequency components, while rapid changes, such as sharp edges, produce high-frequency components. The transform contains a constant coefficient for each aspect in addition to the cosine terms, and this is referred to as being dc, or direct current, while the others are ac, or alternating current. A quantization process is then applied to the coefficients that remove relatively insignificant information. This process is biased towards the characteristics of natural objects (which tend to have gradual changes and hence low frequencies) and to human vision (which is also insensitive to very sharp edges and high frequencies). The quantization step size controls the degree of compression and is relatively small at the low frequencies and large at the high end. Image information is lost in this process, and cannot be recovered by the decoder. Many coefficients are rounded down to zero in quantization, and these are coded as variable length sequences of zeros. The higher-frequency DCT components are more important than the lower ones in determining changes in the image, so some applications may send high and low sets at different cell loss priorities over ATM, and over different logical channels on 3G radio links.

There are two modes of describing the image data: intraframe and interframe. Intraframe is the natural way in which a single frame is compressed on its own; interframe describes successive frames in terms of their differences from the value predicted by the previous frame. Intraframes are called I-frames, and there are two types of predictive interframes, P-frames and B-frames. The majority of these are P-frames, which give the difference to the previous frame, but in order to handle features like the occlusion of one object by another moving object, B-frames are transmitted that have the differences to both preceding and following frames. These are transmitted slightly out of sequence and require buffering and resequencing by the decoder. The buffering causes delay, and as loss of a B-frame does not affect subsequent P-frames, they may sometimes be omitted or sent at a lower network priority. A further type of frame is the D-frame, which is intracoded, but only includes the main dc coefficients and is used for specific applications such as fast-forward.

Most of the early standards organize the pictures into groups of pictures (GOPs) that correspond to repetitive sequences of picture types. A GOP has to contain at least one I-frame and has a length defined as the distance (in pictures) to the next I-frame. A further characteristic parameter is the distance between an I- or a P-frame, which is used as a reference point for further pictures and the next such frame (see Figure 10.1).

In this diagram, the initial dark frame is an I-frame, the clear frames are B-frames, and the shaded frames are P-frames that are used as reference points. The length is 12 in this example. B-frames are not used in many cases, and if absent would be replaced in the diagram by nonreference P-frames. When a picture is transmitted, it is coded on the basis of 8×8 pixel blocks that in turn are organized into macroblocks and either groups of blocks (GOBs) or slices depending on the standard. Division of a picture in this way limits the degree to which errors can be propagated through the picture. A macroblock is a 16×16 pixel area, and the number of coded blocks depends on the (Y, U, V) format, as shown in Figure 10.2 for $(4, 4, 4)$, $(4, 2, 2)$, and $(4, 2, 0)$ formats.

In this figure the first GOB represents luminance, while the second and third are the chrominance components corresponding to horizontal and

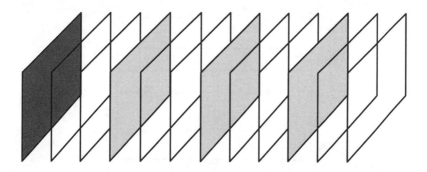

Figure 10.1 Group of pictures.

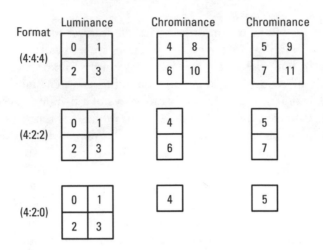

Figure 10.2 Macroblock formats.

vertical scans. The numbering shows which of the 12 blocks comprising a (4, 4, 4) macroblock are retained in (4, 2, 2) and (4, 2, 0) formats. Most potential 3G applications use the (4, 2, 0) format, with six blocks to a macroblock.

A GOB is a complete strip of macroblocks across the picture. A slice is a variable rectangular area, whose position and width need to be specified as in Figure 10.3.

An example of a GOB for the case of H.261 and H.263v1 (see below for these standards) is a row of macroblocks, where the number in the GOB is 22 (equals 352/16) for CIF, with 12 GOBs to a CIF frame or 3 for QCIF.

The stream of compressed video contains header data associated with each of these structures that is vital for interpretation of the stream, and some

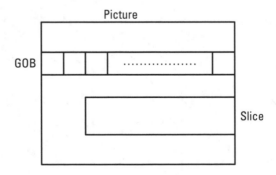

Figure 10.3 GOB and slice.

applications may transmit this over a separate logical channel with higher error protection than the visual content. Early videocompression algorithms were designed for networks with low error rates, and so have little resilience. A few errors in the image have little impact on the viewer, but an error in the video frame header data may make it undecodable, so that it has to be dropped. Many decoders try to conceal these errors by repeating the last frame that was successfully decoded.

Many vendors of compression applications add proprietary features to enhance the basic standard in areas where the standards body has been unable to agree on the best approach, or where the standard is not entirely clear. As a result, best performance is usually obtained where codecs come from the same manufacturer so that the proprietary enhancements can be used.

In general, the low bandwidth of 3G radio access networks means that high degrees of compression have to be used. This is achieved by the use of low-resolution source images (e.g., QCIF) and large quantization steps. Applications that are explicitly designed for mobile networks, such as JPEG-2000 and MPEG-4, also have special features to make them more resilient to transmission errors, as do later versions of H.263. All these applications are treated individually in subsequent sections of this chapter.

UMTS recommends the use of PS streaming services (PSS) [16] that employ RTP/UDP (see Section 7.4.7) for transport of media content, HTTP/TCP for scene description, and text still images and RTSP [19] or HTTP for presentation description. SIP (see Section 7.4.8) controls session initiation, and RTSP, which also makes use of SDP (see Section 7.4.14), provides session control where this is needed. The subsequent sections outline the use of RTP.

10.2.2 JPEG and Motion JPEG

JPEG [20] was originally intended for compression of still images but has been widely used for moving images via its motion JPEG derivative and achieves a typical compression ratio on the order of 15:1. JPEG has one lossless compression mode, the simple predictive differential pulse code modulation (DPCM), in which the differences between predicted and actual values are coded and transmitted, and three lossy modes based on DCT. The three DCT modes differ in the manner of transmission as follows:

- *Sequential:* In this mode blocks are transmitted as soon as they are coded, and the picture appears block by block at the receiver.

- *Progressive:* The original image is successively quantized at several different resolutions with the coefficients stored in a buffer. A full-size picture is received that is initially blurred, but becomes progressively more refined at each stage.

- *Hierarchical:* The image is again quantized at different resolutions, but each pass is used as the prediction for the next layer in a pyramidal approach. The received picture initially appears as a very small size corresponding to the top of the pyramid, and becomes progressively larger at each stage until full size is reached. The sharpness of the images appears roughly the same at each stage.

Most implementations of JPEG, especially those in hardware, use the DCT sequential mode, and this is the only one that is supported in RTP encapsulation for IP transport [21]. The standard RTP header includes a time-stamp based on a 90-kHz clock, and is followed by 8-byte JPEG-specific headers. These headers indicate one of up to 255 JPEG types, a fragment offset, JPEG quantization values, and frame width/height sizes in multiples of 8 pixels.

JPEG is very sensitive to errors, and the burst errors associated with Rayleigh fading are liable to cause areas of corruption in the received image. The bit error rate needs to be less than 1 in 10^6 for reasonable images, so would require use of acknowledged mode in UMTS or assured mode in cdma2000 with multiple retransmissions. JPEG itself is not a real-time application, so the delays associated with retransmission are not a problem, and a low DiffServ code-point can be used in IP.

JPEG is not suited to unnatural images, particularly bilevel diagrams (e.g., black and white), and JBIG is used instead for these, but with similar acknowledged or assured mode parameters.

Motion JPEG produces a sequence of I-frames only, so errors in one frame are uncorrelated with those in subsequent frames. As it is a real-time application, the delays associated with retransmissions are not acceptable, and use of transparent mode in UMTS or unassured mode in cdma2000 will lead to poor quality images, although loss of an occasional frame is not a problem due to use of I-frames only.

10.2.3 H.261 Videoconferencing

ITU-T H.261, or $p \times 64$ as it is also known, is the original video-conferencing standard. It was designed to operate over CS networks using a bandwidth of $p \times 64$ Kbps, where p is an integer in the range 1 to 30. It is based on DCT and the use of interframe prediction, with multiplexing of audio, video, data, and signaling information before transmission. The normal frame rate is 30 per second, but the standard allows up to three frames to be interpolated between transmitted frames, thereby reducing the rate to 15, 10, or 7.5 in order to cut the required bandwidth. At high bandwidth the source image is CIF, but for low rates QCIF is used with format (4, 2, 0) in either case. Groups of four luminance and two chrominance blocks are combined to form a macroblock. The standard allows for the limited degree of movement likely in a videoconference, and has an optional motion vector per macroblock to support this aim. The standard makes use of eight different modes of coding the macroblocks as follows:

1. Intercoded;
2. Motion compensated (MC);
3. Non-MC;
4. Intracoded;
5. Not coded (where none of the transformed component blocks has significant energy);
6. Intercoded plus quantization step size definitions;
7. MC plus quantization step size definitions;
8. Intracoded plus quantization step size definitions.

H.261 does not use B-frames on account of the delay that they introduce and their limited value in the videoconferencing environment.

If a macroblock is corrupted by bit errors during transmission, the corruption will persist until the next intracoded version of that macroblock is received. H.261 specifies that an individual macroblock must be updated at least once every 132 frames. In the worst case of the low bit rates appropriate to mobiles, this could mean a corrupted image persisting for up to 17.6 seconds at the lowest frame rate of 7.5 per second. At these low frame rates it should be possible to allow up to one retransmission on the radio access network, but at typical BER of 10^{-2} or 10^{-3} many of the macroblocks will still be corrupted, so received video images will show persistent disfigurements. H.261 includes an option for forward error correction on an end-to-end basis between codecs using a BCH block code; despite the use of FEC on the radio links, this option is likely to be beneficial.

At the lowest data rates of 64 and 128 Kbps, the large associated quantization step leads to a blocky appearance of the received images. This has two characteristic features: a high-frequency noise called mosquito noise, associated with blocky areas in variable positions, and staircase noise, where the blockiness breaks up a sharp edge. These defects can be reduced with the use of filters, but at the expense of slight blurring.

The quantity of information to be transmitted varies according to the type of macroblock, particularly intracoded versus intercoded, and this causes a problem on CS or other CBR network links. H.261 uses buffering to compensate for this, but at the expense of adding both delay and jitter, together with some bit dropping or stuffing. The latter technique may not be equally applied over a complete frame and can lead to image quality variability over the frame as a result; in addition, it is not fully standardized, so different H.261 implementations may not be compatible in this aspect of rate control.

The speech component is based on either 64-Kbps PCM or 32-Kbps ADPCM and so requires transcoding for mobiles, leading to delay and loss of quality.

A 3G receiver of H.261 images is thus likely to encounter an overall poor image quality due to the deficiencies of the algorithm at the lowest transmission rates, together with frequent disfigurements of parts of the image for several seconds at a time due to the errors resulting from the radio access network. The residual BER must be better than 10^{-6}.

Although H.261 was designed for CS networks, it is also the main video compression standard recommended in H.323 (see Section 7.5) for multimedia over PS networks, where it is encapsulated in RTP. There is a standard payload format when transported by RTP [22]. This has a special H.261 header that includes information on the encoded video frame including a flag to distinguish I-frames, a flag to show if motion vectors are used by the session, a GOB number (relative to the start of the frame), a quantization value, macroblock address predictor, and when motion vectors are in use, the horizontal and vertical motion components. Packet loss or corruption is a major problem for this application over mobile networks.

10.2.4 MPEG-2

The first widely deployed standard for compression of video images with a large degree of movement was MPEG-1, which was intended for compression of audio-video for storage purposes, particularly with CD-ROM drives. As a result, it lacks resilience to errors and is not suited to network transmission.

MPEG-2 [12], which is backwards compatible with MPEG-1 and has been adopted by the ITU as ITU-T H.262, was designed to be usable over networks and to support many additional features associated with transmission of TV programs and video on demand, such as distribution of several distinct programs simultaneously. Consequently, it carries both a program stream and a transport stream.

The program stream is able to multiplex several different programs together through the transmission of successive program packs. The individual program packs contain features such as a program code, table of contents, security, access control, and some audio-video characteristics. It is transmitted in the form of relatively large packets.

The transport stream carries the actual program material, and is transmitted as 188-byte packets. These packets each have a 4-byte header and 184 bytes of payload, with the size chosen so that one 188-byte packet would fit directly into four ATM AAL1 cells (each with 1 control byte and 47 traffic bytes). The standard 4-byte header contains a program identifier, synchronization byte, error flag, start indicator, and transport priority level. It can optionally be extended at the expense of the payload field to include a 90-kHz program reference clock (PCR) time-stamp to allow the receiver to reduce jitter. Use of these time-stamps is a more effective way of controlling jitter than the end-to-end use of AAL1, so ATM AAL5 or AAL2 can be used instead with PCR. If AAL5 is used, then one, two, or more transport stream packets are concatenated and fitted into five, eight, or more ATM cells, with some bytes left over that may be used in a proprietary manner.

MPEG-2 has to provide compression of both audio and video material (its MP1–3 audio compression standards have already been described earlier in this chapter in relation to the ubiquitous MP3 application). A more efficient audio compression technique, advanced audio coding (AAC), is also used in later implementations. The video compression component is closely related to that of JPEG, but with allowance for motion and transmission errors. It again uses DCT block transforms that are optimized for natural objects and human perception, and, like H.261, uses both intercoded and intracoded objects, in this case complete frames. The intercoded frames are both P-frames and B-frames as mentioned in the introductory section of this chapter. An I-frame is sent every 12 frames, so errors do not persist nearly as long as with H.261.

MPEG-2 covers an extremely wide range of applications; its functionality is described in terms of five profiles and four levels. The profile defines a subset of MPEG-2 functionality, and the level describes a constraint on the parameter values. These are summarized in Tables 10.11 and 10.12.

Table 10.11
MPEG-2 Profiles

Profile	Basic Characteristics
Simple	Does not use B-frames, mainly for software decoders
Main	Common low-cost implementation with I-, P-, and B-frames
Spatial	Main with added spatial scalability
SNR	Spatial with added SNR scalability
High	SNR with 4, 2, 2 format

Table 10.12
MPEG-2 Levels

Level	Source Bit Rate (Mbps)	Quality (pels × fps)	Typical Application
Low	4	352 × 288 × 30	VHS-grade video
Main	15	720 × 476 × 30	Broadcast quality
High 1440	60	1,440 × 1,152 × 30	HDTV
High	90	1,920 × 1,152 × 30	Film production

MPEG-2 has typical compression ratios of 15–30:1, so with a maximum transmission rate of 2 Mbps only two levels are relevant for 3G networks. The relevant levels are low and main with compressed rates of 1 Mbps or less.

The most common MPEG applications are main for both profile and level. Main is used for both storage media and broadcasts. It sends all pictures, provides no scalability, and uses the (4, 2, 0) format.

The second applicable profile is SNR scalable, which is more important for networks. SNR refers to the fact that this profile first quantizes a frame, then decodes this output and compares it with the original frame. The differences between the original and decoded version are then compressed and sent as a set of enhancements alongside the standard code. SNR scalable is aimed at ATM networks, and the enhancements will typically be sent at a different cell loss priority to the standard code so that in the event of network congestion or restricted bandwidth links they will be preferentially dropped. Where both the standard and enhanced codes are received, a decoder will normally combine both signals to form an output that is free from quantization noise.

Scalable in this instance refers to the option for dropping the enhancements, but other MPEG-2 profiles have additional scalability features to allow interoperability of equipment and networks using different image sizes and frame rates. Simple profile is often used for videoconferencing because of the need for minimal delay.

MPEG-2 describes motion through the use of a macroblock identity and average motion vector per macroblock in P-frames; while B-frames use two motion vectors, one for time advance and the other for time retard.

MPEG-2 has several options for handling errors in addition to the use of relatively short packets. The first of these is an option for forward error correction and another is for data partitioning. Data partitioning is an extension of SNR scalability and is based on two priority levels. The basic code can be divided between the two levels so that less important information is sent at the same level as the enhancements. Typically, the lesser information would consist of the higher-frequency DCT coefficients to which the human eye is less sensitive. The more sophisticated of the decoders can also implement error concealment by replacing defective data with interpolated data, although this can sometimes turn out to be counterproductive.

Although MPEG-2 was designed for use over ATM, it is frequently encapsulated in IP using RTP. Audio and video components of the transport stream are separately encapsulated (that for the audio part was described in Section 10.1.3 for MP3). The RTP header includes the payload type (distinct from that for audio), a video stream identifier, and a 32-bit presentation time-stamp to help the network control jitter. This is followed by a 4-byte video-specific header that includes a 10-bit temporal reference for the picture within the current GOP, the picture type (I-, P-, B-, or D-), and various flags. The MPEG-2 RTP payload types are 33 for video and 14 for audio.

Several other well-known standards for video compression and storage are substandards of MPEG-2; these include DVD, Video-CD, and CD-i. They have specialized restrictions on parameters and require gateways for interworking with MPEG in general.

10.2.5 JPEG-2000

This standard was devised to be free of the limitations of the original JPEG and to be suitable for use over mobile phone networks. Thus, it is capable of handling synthetic images as well as natural ones and has resilience features for error-prone networks. It is recognized that even with the aid of forward error correction on the radio links the error rate may still be excessive, so resilience is provided at the source coding stage also.

The features that provide resilience are hierarchical resynchronization after errors, data partitioning, and error concealment by the decoder. Unlike the earlier JPEG, it does not use the DCT transform, but employs the newer wavelet transform instead. The input is first decomposed by the wavelet transform into frequency subbands that are then split into bit-planes and then into blocks. Each block is coded independently to prevent error propagation.

Segments from various blocks are collected into a packet, and the packet is provided with a resynchronization header. The encoder can choose the number of blocks to go into a packet and, if it has feedback on the error rate, can decide this so as to get the best performance. If the error rate is high, the number of blocks per packet is reduced at the penalty of a higher overhead, while at low error rates the number is increased. Each packet carries a resynchronization marker (RM) and an identification number; the latter identifies the subband, SNR layer (analogous to MPEG), and spatial partition to which the packet belongs. After an error, all data is discarded until the next RM is received.

The different bit-planes can be assigned different qualities of service so that in congested networks the least important parts of the image can be preferentially dropped.

10.2.6 MPEG-4

This standard [23–27] is explicitly designed to meet the problems of sending moving images and videoconferencing over low bit rate, error-prone networks such as the PSTN and mobile networks.

Initially it was expected that only low resolution images, such as QCIF and sub-QCIF, would be supported on account of the low bandwidth, but the success of the standard has led to its use at higher speeds for much higher quality sources, such as 4CIF and 16CIF, which have 4 and 16 times the number of pixels as CIF, respectively.

MPEG-4 works with audiovisual objects (AVO) together with a scene description procedure for assembling them into the full display rather than transmitting complete frames as a whole. AVOs can either be primitive objects, such as the sound of somebody talking or the image of a person, or else compound objects, such as a person talking. An instance of a visual object at a specific time is called a video object plane (VOP). Binary format for scenes (BIFS) is the scene description language, which in addition to reducing latency also provides a quick means of updating scenes.

Like MPEG-2 it has multiple profiles and levels, the main difference being the addition of the QCIF and sub-QCIF levels. The profiles determine which of the various MPEG-4 object types are supported, while the levels indicate the constraints placed on parameter values. The feature types in MPEG-4 are as follows: (1) basic I- and P-VOP, (2) error resolution, (3) short headers, (4) B-VOP, (5) temporal scalability of P-VOP, (6) binary shapes, (7) sprites, and (8) N-Bit. The related profile types for natural images are shown Tables 10.13 and 10.14 [28] where "y" indicates support.

Table 10.13
MPEG-4 Profiles

Profile Name	Object Types							
	(1)	(2)	(3)	(4)	(5)	(6)	(7)	(8)
Simple	y	y	y					
Simple scalable	y	y		y				
Core	y	y	y	y	y	y		
Main	y	y	y	y	y	y	y	
N-bit	y	y	y	y	y	y		y

Table 10.14
MPEG-4 Levels

Profile Name	Level	Bit Rate	Maximum Objects	Image Format
Simple	L0	28.8 Kbps	2	Sub-QCIF
Simple	L1	64.0 Kbps	4	QCIF
Simple	L2	128.0 Kbps	4	CIF
Simple	L3	384.0 Kbps	4	CIF
Core	L1	384.0 Kbps	4	QCIF
Core	L2	2.0 Mbps	16	CIF
Main	L2	2.0 Mbps	16	CIF
Main	L3	15.0 Mbps	32	R 601
Main	L4	38.4 Mbps	32	$1{,}920 \times 1{,}088$

There are also several profiles for specific artificial features, notably types of animation, and a hybrid profile that has both the natural and synthetic features.

The basic technique for coding the images is essentially that of JPEG-2000 using wavelet transforms, and one of the key features is the use of VOPs. A video frame is described as a layered structure in which each main object has its own VOP, with the VOPs superimposed on each other and on a background field. For example, in the broadcast of a game of baseball or cricket, the main features, such as the ball and key players, could have their own VOPs (but depending on profile and level), while the sports ground would be the background VOP or sprite. Each VOP is treated independently and its evolution depicted through its own sequence of I-, P-, and B-frames. The background VOP is treated as a sprite in some profiles and coded much less frequently (with information on camera angles used to update the display) than the main moving objects, and so it contributes much less network traffic in MPEG-4 than in MPEG-2 or H.263. Packets belonging to the different VOPs can be treated as different logical channels over the network and be provided with differing qualities of service if required. This applies particularly in video-conferencing and films for multilingual presentation. A key factor for ease of user interpretation in each case is lip synchronization of the speakers—so lips can be given their own VOP to be superimposed on that of the face and be provided with a higher degree of error protection on the radio link. In the case of the multilingual films, separate VOPs for the lip movements can be dubbed in addition to the sound to provide a much more convincing presentation than traditional sound dubbing. Use of the MPEG time-stamps for the lip VOPs and the audio stream allows precise lip synchronization.

Another major feature of MPEG-4 is the use of visual texture coding (VTC) to provide high-quality synthetic images. VTC uses different wavelet transform and quantization techniques to the main video module, and differs in its resilience features also.

MPEG-4 has three sets of audio procedures as follows:

1. Natural sound can be sampled at up to 8 kHz and parameter coded by harmonic vector excitation coding (HVXC) at bit rates from 1.2 to 4 Kbps. These algorithms have low MOS ratings of 3 or less.

2. CELP algorithms can also code natural sound with sampling rates in the range 8 to 16 kHz and bit rates of 6 to 24 Kbps. The precise algorithms differ from those used for the ITU G. Series of recommendations and those for mobile networks.

3. High-quality sound can be provided using AAC algorithms with sampling rates from 8 to 96 kHz and bit rates from 16 to at least 64 Kbps.

Audio scaling is possible by using a CELP base with AAC enhancements.

MPEG-4 has a much higher degree of end-to-end resilience than MPEG-2 to improve operation on mobile networks. One option is for the encoder to insert RMs much more frequently, variously by dividing the picture into video packets made up of an integral number of consecutive macroblocks spanning more or less than one row, and inserting RMs every Z bits. Z is not standardized, but the value 512 is sometimes used, which allows short burst errors to be confined to a few macroblocks in high-activity areas of the picture instead of affecting a whole row of macroblocks.

A header extension code is another important error protection feature. A header that describes the spatial aspects of the image, the time-stamps, and the mode precedes each video frame. If this is corrupted, the picture cannot be decoded, so the header extension is used to repeat the header to avoid that, particularly with the VTC part.

Data partitioning is the next technique to reduce the amount of data lost due to an error. A motion boundary marker (MBM) is inserted between the Motion and VTC parts of a packet to enable an error to be localized to one part or the other. Data in the defective part is then discarded until the next applicable RM. (See Section 10.2.7 for an example of RM and MBM use, in relation to H.263, which also does this.)

The motion part, but not VTC, also uses reversible variable length codes (RVLC), whose format is such that they can be read either forward or backward. If a decoder finds an error while decoding in the forward direction, it can go on to the next RM, then decode backward until it comes to the error. The RVLCs are longer than would otherwise be normal and so require extra bandwidth, but this disadvantage is more than offset by the extra resilience on radio links.

One error correction feature that is not used in MPEG-4 is end-to-end FEC, as this is not required for high-quality network sections, such as optic-optic links, but it is assumed to be provided by the network itself on poor quality parts, such as the radio links.

MPEG-4 is designed to be independent of the underlying transport network, unlike MPEG-2, which was designed for use with ATM. AVOs are sent in one or more elementary streams, each of which contains an object descriptor that includes a QoS requirement. There is then a three-layer architecture for processing these elementary streams, as shown in Figure 10.4.

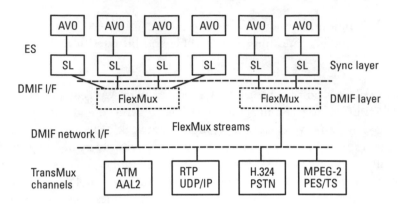

Figure 10.4 DMIF layers.

The synchronization layer (SL) is mandatory and packetizes the elementary stream (ES) into access units to which it adds time-stamps to allow synchronization at the receiver. A typical implementation provides one ES for each AVO, with potentially an additional ES for the BIFS scene description commands, so the simple profile applications most likely to be used over mobile networks would not have more than five ES. SL interfaces to the delivery multimedia integration framework (DMIF) [29] and a FlexMux layer that is optional but multiplexes different types of access unit into Flex-Mux streams with common needs. Multiplexing is most likely to be required for the more complex implementations where there are many objects and hence multiple ES differing little in their QoS needs. The FlexMux output streams have specific QoS needs and are sent to the TransMux channel for selection of an appropriate network traffic channel using a standard DMIF network interface, but these traffic channels are not part of the specification. DMIF contains its own signaling component, which includes specification of its QoS needs, but many applications may just use RTSP [19] with a fixed network interface despite RTSP's inability to specify QoS features other than bandwidth and packet size. Typical network channels are RTP/UDP/IP, AAL2/ATM for audio, H.324/PSTN, or the original MPEG-2 ATM streams. In the case of RTP, there is a unique RTP session for each FlexMux (or in its absence ES) stream.

The IETF has defined a set of procedures for carrying MPEG-4 in RTP packets [30]. The video payload type is usually selected out of band from the dynamic range (96–127) by H.245 or SIP commands, or defined independently for specific MPEG-4 applications. MPEG-4's combined configuration/ES mode is used with RTP, and the bit stream is inserted directly into RTP packets with only the standard RTP header. Unlike H.261 and MPEG-2, there is no need for a specialized header as MPEG-4 has sufficient intrinsic resilience to avoid this requirement. The only exception to this occurs with MPEG-4's short header option, and if this is used, then the extra header recommended for H.263 [31] (and below) is proposed. Ideally, an RTP packet should only contain a single VOP, or fragment thereof, but with very small VOPs an exception may be made. In that case, the RTP time-stamp refers to the first VOP, and the others have to be calculated by the decoder. When MPEG-4 data is placed in RTP packets, the various MPEG headers should follow the standard RTP header. On networks with a high probability of packet loss, such as wireless links, a complete video packet (i.e., VOP headers plus VOP) should be placed in a single RTP packet if possible so that its loss does not affect interpretation of other packets. This may require adjustment of the video packet sizes for compatibility with the IP MTU.

The audio stream is treated differently and must use the low overhead MPEG-4 audio transport multiplex (LATM) tool with RTP [30]. It is recommended to have one audiomux element per RTP packet if possible. The payload type is again dynamic, and different payload types have to be used for each level in scalable audio. Multiplexing of objects and of scalable audio layers has to be avoided in LATM, which is only for natural audio, when used with RTP.

MPEG-4 is considered as one of the most important real-time applications for use over UMTS networks. Applications over UMTS are most likely to use bit rates in the range 28 to 144 Kbps. The residual BER needs to be about 10^{-6} to avoid any visible degradation [32], with flaws becoming increasingly apparent until a usability limit of 10^{-3} or 10^{-4}. The exact limits will depend on how fully MPEG's own error protection mechanisms are implemented. There may be an RAB subflow in the UTRAN for each MPEG-4 ES; however, at the low bit rates for most mobile users there will be a maximum of 4 objects, and hence not more than that number of subflows, with only one in many instances. UMTS delay targets are in the region of 150 to 400 ms, with an expectation that the codecs will contribute about half of this.

10.2.7 H.263, H.324, and 3G-324M

The ITU recommendation for supporting multimedia telephony over low bit rate CS networks is H.324. This draws on four other recommendations for its component functions as follows:

1. H.245/248 for call control;
2. G.723.1 for audio compression;
3. H.261/263 for moving images and videoconferencing;
4. H.223 for multiplexing the signaling, data, audio, and video channels.

Of these, H.248 was described implicitly in relation to Megaco Protocol [33] in Section 7.4.12, while H.723 was described in Section 10.2. The 3G UMTS standard for audio codecs is AMR, not G.723.1, so the 3G Partnership Program has a modified version of H.324 called 3G-324M [34] that incorporates this change and a few others. The corresponding ITU recommendation for PS networks is H.323 (see Section 7.5 and [35]), but this is not directly supported in UMTS.

Both H.223 and H.263 were originally intended for operation on networks with low-speed links and fairly low error rates, such as PSTN and N-ISDN, but have since been updated to allow for the high error rates on mobile networks. H.223 originally had only one style of multiplexing, which is now referred to as level 0, but three additional levels were added for error handling with the following features:

- Level 0 provides variable length packets with a 1-octet header defining the logical channels carried and uses standard 01111110 HDLC flags with bit stuffing as frame delimiters.

- Level 1 is intended for low error rate networks and is similar, but uses 16-bit flag frame delimiters without bit stuffing instead of HDLC flags to reduce the chances of flag corruption.

- Level 2 is for moderate error rates and differs from level 1 by the use of a Golay code to provide error protection for the packet header.

- Level 3 is for high error rates and differs from level 2 by provision of error protection for the payload field by any of several techniques (ARQ, CRC checks, or a rate compatible punctured convolution code).

H.263 has had three versions so far, with the original having been designed for CS networks, and changes made later for the benefit of mobile networks in particular. Version 1 of H.263 video compression is fairly similar to the earlier H.261, but with better motion compensation and a different coding for the transform vectors [34]. Like MPEG-4 it was designed for small format images at low speeds but is also used for larger formats, such as CIF and 4CIF, at higher speeds. H.263 has one motion vector per macroblock of 16 × 16 pixels and bases this on predictions from each of three adjacent macroblocks. Like H.261 but unlike MPEG-4, horizontal rows of macroblocks are grouped into GOBs for transmission and numbered from top to bottom in a vertical scan (unlike H.261). Version 1 only had one (optional) method of error handling, which was to send H.245 messages back from the decoder to the encoder to request specific macroblocks in the current image to be coded in intra mode to limit propagation of errors. Because of the network delays in receiving the decoder feedback and the effects of motion, the encoder needs a lot of buffering to track the error. In version 2, often referred to as H.263+, an additional three options for error handling were introduced:

- *Slice structure method:* This uses slice structure codes (SSC) throughout the bit-stream for resynchronization analogously to MPEG-4 (Figure 10.5), where a slice refers to a group of blocks.

- *Independent segment decoding:* The above method does not allow for the effects of motion, so this technique prevents errors propagating to neighboring slices by restricting the motion compensation vectors to lengths that keep them within their original segment. This causes lower compression owing to increased differences between successive pictures.

- *Reference picture selection:* This is another option that relies on feedback from the decoder for error limitation. In this case the decoder requests one of several correctly received and decoded frames to be used as the reference in intercoded pictures instead of an erroneous one. This requires considerable codec memory.

- *Video redundancy coding (VRC):* This is another option that can be implemented in conjunction with reference picture selection for error-prone PS networks. Instead of basing sequences of P-frames on a unique reference picture, multiple threads can be defined, each of which is based on a separate reference picture. The effect of an error in one reference picture is then confined to only one thread. Periodically a common synchronization frame is sent on all threads.

H.263 uses several H.245 features for support of video-feedback as follows:

- *Freeze frame:* picture frozen until release or a 6-second time-out;
- *Fast update request:* request for intracoding of macroblocks for error limitation;
- *Freeze frame release:* back to normal transmission after either of the two preceding options;
- *Continuous presence multipoint:* a multipoint control unit that can transmit up to four independent video streams for split-screen presentation in a single H.263 video stream.

H.263+ offers three scalability options: spatial, temporal, and SNR. Spatial scalability typically involves two sets of pictures, basic and enhanced, with the latter sent at twice the resolution in each plane (e.g., CIF instead of QCIF). Temporal is given by two frame rates offered by either inclusion or exclusion of B-frames. SNR is similar to that for MPEG with an optional set of quantization errors. In an IP environment, distinct IP addresses and port numbers are used for the different options so that a user can easily log on to the appropriate quality.

H.263+ pictures are transmitted using a multiple layer structure with different headers for each layer as shown in Figure 10.5 [36].

The meanings of the uninterpreted acronyms in this diagram are as follows:

- CPM stands for continuous presence multipoint and is a flag that indicates that four independent video bit-streams may be sent

Layer	Header fields with sizes (bits)					
Picture	Start code (22)	Temp ref (8)	Picture type (13)	Pict (5) quantzn	CPM (1)	PEI (1)
Group of blocks	Start code (17)	GOB nom (5)	GOB ID (2)	GOB quantzn	GBSI	
Macro block	COD (1)	MCBPC (var.)	CBPY (var.)	DQUANT (2)	Coded motion vectors	
Block	INTRADC (8)	TCOEF (coded DCT coefficients)				

Figure 10.5   H.263 frame layers.

within a single H.263 video channel—if so, the field GBSI is used in the GOB header.

- PEI is a flag that indicates the presence of an optional field.

- GBSI is the GOB substance indicator for the streams in the multipoint case indicated by the CPM flag.

- The COD flag is used to show whether the macroblock has been coded, or skipped because it has not changed significantly.

- MBCPY is a variable length field that describes the type of macroblock and which chrominance blocks are coded.

- The CBPY field shows which of the four luminance blocks within the macroblock have been coded (some may have been skipped for lack of change).

- DQUANT shows the change in quantization parameters for the macroblock since the previous one, and for the first macroblock is set to the default quantization parameter (QP).

- The block layer shows the coding of the actual video data, where INTRADC represents the constant coefficient in the DCT expansion for an I-frame, and TCOEFs describe the coefficients for the cosine terms. Some operations, such as fast-forward, only use INTRADC.

If slice structure mode is used in H.263+, then a slice structure header is used instead of the GOB header. This includes a specification of the slice size in addition to comparable material to that for the GOB.

H.263+ also uses resynchronization markers and motion boundary markers for data partitioning as shown in Figure 10.6.

This shows how the macroblock is split into parts for motion vectors and picture data. The number of macroblocks following the RM is provided, followed by the default QP, followed in turn by the motion data for each successive macroblock. After the MBM, the data headers for each macroblock are given, with the actual DCT data for the individual macroblocks behind the set of headers.

H.263 version 3 (or H.263L) is intended to provide still better performance on low bit rate links and on poor quality networks. It is also intended to have a much greater compatibility with MPEG-4.

A further improvement to H.263 is the recent addition of profiles to classify sets of functionality (H.263 Annex X) for version 2. Those most relevant to UMTS are as follows:

Use of resynchronization marker and motion boundary marker

RM	MB#	QP	Motion data	MBM	DCT data

where the form of motion data and DCT data for H.263 is:

COD	MCBPC	Motion vectors	COD	MCBPC	Motion vectors	COD ....

CBPY	DQUANT	CBPY	DQUANT	DCT data	DCT data ....

Figure 10.6 Data partitioning.

- The baseline profile (profile 0) stands for H.263 with no optional modes of operation. It includes the basic coding tool set common in modern video coding standards. It provides simple means to insert resynchronization points within the video bit-stream, and therefore, it enables recovery from erroneous or lost data. This has several levels, and UMTS specifies level 10 [16].

- The version 2 interactive and streaming wireless profile (profile 3) provides enhanced compression efficiency when compared to the baseline profile and is also recommended in [16]. Moreover, it provides enhanced error resilience for delivery to wireless devices. Specifically, profile 3 includes the following optional coding modes:

  - *Advanced INTRA coding (Annex I):* Use of this mode improves the compression efficiency for INTRA macroblocks (whether within INTRA pictures or predictively coded pictures);

  - *Deblocking filter (Annex J):* A deblocking filter improves image quality by reducing blocking artifacts. When compared to deblocking filtering performed as a postprocessing operation, the deblocking filter mode reduces the amount of required memory, as no additional picture memory is needed for the filtered images. This mode also includes the four-motion-vector-per-macroblock feature and picture boundary extrapolation for motion compensation, both of which can further improve compression efficiency.

  - *Slice structured mode (Annex K):* This adds resynchronization markers (see Figure 10.6).

- *Modified quantization (Annex T):* This mode enables flexible quantizer control that can be used in sophisticated bit rate control algorithms. In addition, it improves chrominance fidelity.

RTP Encapsulation

RFC 2190 [37] was originally defined for RTP encapsulation of H.263, but the latter RFC 2429 [31] is recommended for H.263+, and it is the latter that is outlined here. The video bit-stream is directly packetized within RTP with the exception of two null bytes at the beginning of some of the start codes. The standard RTP header is used with the following principles:

- The RTP marker bit is set to 1 when the packet contains the end of the current frame, and otherwise is set to 0.
- The payload type indicates H.263+.
- The RTP time-stamp refers to the sampling time of the first video frame in the RTP packet. If a video frame covers several RTP packets, then they all carry the same time-stamp. For spatial and SNR scalability, packets referring to the same frames at different scales carry the same time-stamps, but temporal scalability requires special ordering to handle the B-frames.

The standard RTP header is followed by a specific H.263+ header, which is of variable length, to describe the details of the different encoding options in use and the possible inclusion of an extra picture header. The basic structure is shown in Figure 10.7.

The significance of these fields are listed below:

- *RR:* These 5 bits are reserved and set to 0.

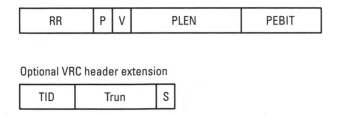

Figure 10.7 H.263+ RTP header.

- *P:* This 1-bit flag indicates the start of a picture, GOB, slice, or video sequence end. Two zero bytes have to be prefixed to the start codes in the payload to reconstruct the H.263+ start codes.

- *V:* This 1-bit flag indicates the presence of an 8-bit VRC field at the end of the standard H.263+ header.

- *PLEN:* This 6-bit field indicates the length of an extra picture header if it is present; otherwise it is set to 0.

- *PEBIT:* This 3-bit field indicates a number of bytes to ignore at the end of the extra picture header if present.

- *VRC field:* This optional extension includes a 3-bit thread ID (allowing up to seven threads), a 4-bit packet number within the thread, and an *S*-bit that is set for a synchronization frame.

3G-324M also includes optional support for MPEG-4 visual simple profile at level 0 and makes recommendations for how visual object start codes should be aligned with the start of SDUs.

Cdma2000 recommends the use of H.324 or MPEG-4 visual for videoconferencing but does not specify the speech codec component. The image sizes recommended are CIF and QCIF.

H.263 is also an optional form of coding for use with H.323 packet mode multimedia; some of the issues for this in UMTS were discussed in Chapters 7 and 8.

Early versions of H.263 need residual BER of less than 10^{-6}, but later versions with better error correction can stand higher rates of up to 10^{-4} as with MPEG-4.

When 3G-324M runs over UMTS, it will require the three logical channels used for AMR speech (see Section 10.1.1) together with at least one for the video image, so the number of RAB subflows and TFCS components will each be at least four.

References

[1]   Goldberg, R., and L. Rick, *A Practical Handbook of Speech Coders*, Boca Raton, FL: CRC Press, 2000.

[2]   IETF RFC2658, "RTP Payload Format for Qualcomm Pure Voice (QCELP)," 1999.

[3]   ETSI TS 126 102 "AMR Codec," V.4.1.0 (2001), http://www.etsi.org.

[4]   ETSI TS 126 103, "List of Speech Codecs," V.4.1.0 (2001), http://www.etsi.org.

[5]   ETSI TS 123 153, "Out of Band Transcoder Control," V.4.3.0 (2001), http://www.etsi.org.

[6]   ETSI TS 128 062, "Tandem Free Operation of Speech Codecs—Service Description," V.4.1.1 (2001), http://www.etsi.org.

[7]   ETSI TS 126 115, "Transmission Delay and Echo Control Planning for Speech and Multimedia Services," V.4.0.0, 2001, http://www.etsi.org.

[8]   ETSI TS 26.101, "Adaptive Multirate Speech Codec—Frame Structure," V.4.1.0 (2001), http://www.etsi.org.

[9]   ETSI TS 126 235, "Packet Switched Conversational Multimedia Applications: List of Default Speech Codecs," V.4.0.0 (2001), http://www.etsi.org.

[10] Third Generation Partnership Project, TS 26.201, "AMR Wideband Speech Codec—Frame Structure," http://www.3gpp.org.

[11] Frame Relay Forum, FRF.12, http://www.frforum.com.

[12] ISO/IEC International Standard 13818, "Generic Coding of Moving Pictures and Associated Audio Information," 1994.

[13] IETF RFC2250, "RTP Payload Format for MPEG-1, MPEG-2," 1998.

[14] IETF RFC3119, "A More Loss-Tolerant RTP Payload Format for MP3," 2000.

[15] IETF RFC2046, "Multipurpose Internet Mail Extensions; Part Two, Media Types," 1996.

[16] ETSI TS 126 234, "End-to-End Transparent Streaming Service; Protocols and codecs," V.4.1.0 (2001), http://www.etsi.org.

[17] "Progressive Bi-Level Image Compression, Rev 4.1," ISO/IEC/JTC1/SC2/WG9, CD11544, 1991.

[18] Ghanbari, M., *Video Coding*, New York, IEEE, 1999.

[19] IETF RFC2326, "Real Time Streaming Protocol (RTSP)," 1997.

[20] Pennebaker, W. B., and J. L. Mitchell, *JPEG Still Image Compression Standard*, New York: Van Nostrand Reinhold, 1993.

[21] IETF RFC2435, "RTP Payload Format for JPEG," 1998.

[22] IETF RFC2032, "RTP Payload Format for H.261," 1996

[23] ISO/IEC 14496-1, "Information Technology—Coding of Audio-Visual Objects—Part 1: Systems," 1999.

[24] ISO/IEC 14496-2, "Information Technology—Coding of Audio-Visual Objects—Part 2: Visual," 1999.

[25] ISO/IEC 14496-3, "Information Technology—Coding of Audio-Visual Objects—Part 3: Audio," 1999.

[26] ISO/IEC 14496-2, "Information Technology—Coding of Audio-Visual Objects—Part 2: Visual, Amendment 1: Visual Extensions," 2000.

[27] ISO/IEC 14496-3, "Information Technology—Coding of Audio-Visual Objects—Part 3: Audio, Amendment 1: Audio Extensions," 2000.

[28] Home, C., A. Puri, and P. K. Doenges, "MPEG-4 Visual Standard Overviews," in *Multimedia Systems, Standards and Networks*, A. Puri and T. Chen (eds.), New York: Marcel Dekker, Inc., 2000.

[29] ISO/IEC 14496-6, "Information Technology—Coding of Audio-Visual Objects—Part 6: Delivery Multimedia Integration Framework," 1999.

[30] IETF RFC3016, "RTP Payload Format for MPEG-4 Audio/Visual Streams," 2000.

[31] IETF RFC 2429, "RTP Payload Format for H.263+," 1998.

[32] ETSI TS 123 107 "QoS Concept and Architecture," V.4.3.0 (2001), http://www.etsi.org.

[33] IETF RFC3015, "Media Gateway Control Protocol (Megaco)," 2000.

[34] ETSI TS 126 111, "Codec for Circuit Switched Multimedia Applications: Modifications to H.324," V.4.0.0 (2001), http://www.etsi.org.

[35] ITU H.323, http://www.h323forum.org/papers.

[36] Ngan K. N., C. W. Yap, and K. T. Tan, *Video Coding for Wireless Communication Systems,* New York: Marcel Dekker, Inc., 2001.

[37] IETF RFC2190, "RTP Payload Format for H.263," 1997.

List of Acronyms

1G   first generation

2G   second generation

3G   third generation

3GPP   Third Generation Partnership Project

3GPP2   Third Generation Partnership Project Two

AAA   authentication, authorization, and accounting

AAC   advanced audio coding

AAL   ATM adaptation layer

ABR   available bit rate

ABT   ATM block transfer

ACELP   algebraic code excited linear prediction

ACH   access channel

ACK   acknowledgment

ADPCM   adaptive differential pulse code modulation

AF   assured forwarding

AM   acknowledged mode

AMPS   Advanced Mobile Phone System (analog)

AMR   adaptive multirate (codec)

AMR-WB   adaptive multirate wideband AMR

ANM   answer message

APN   access point name

ARQ   automatic repeat request; admission request

ATC   ATM transfer capability

ATM   asynchronous transfer mode

AUC   authentication center

AVO   audio-visual object

BCCH   broadcast control channel

BCH   Bose, Chaudhuri, and Hocquenghem; broadcast channel

BER   bit error rate

BGCF   breakout gateway control function

BICC   bearer independent call control

BIFS   binary format for scenes

B-ISUP   broadband integrated services user part

BLER   block error rate

BMC   broadcast and multicast control

BSAP   base station application part

BSC   base station controller

BSMAP   base station management application part

BSN   block sequence number

BSS   base station system

BSSGP   BSS GPRS Protocol

BTS   base transceiver station

BVC   BSSGP virtual connection

BVCI   BVC identifier

CACH   common access channel

CAM   channel assignment message

CAMEL   customized applications for mobile enhanced logic

CBR   constant bit rate

CCCH   common control channel

CCTrCH   coded composite transport channel

CDG   CDMA Development Group

CDMA   code division multiple access

CDT cell delay time

CDV cell delay variation

CELP code excited linear prediction

CHAP Challenge Handshake Authentication Protocol

CID context identifier; channel identifier

CIF common intermediate format

CMIP Common Management Information Protocol

CNAME canonical name

COPS common open policy server

CPCH common packet (transport) channel

CCPCH common control physical channel

CPICH common pilot channel

CPS common part sublayer

CRC cyclic redundancy check

CS coding scheme

CS-ACELP conjugate structure-algebraic code excited linear prediction

CSCF call state control function

CSRC contributing source

CWND congestion window

DAMPS Digital Advanced Mobile Phone System

DCCH   dedicated control channel

DCH   dedicated (transport) channel

DCT   discrete cosine transform

DHCP   Dynamic Host Control Protocol

DL   downlink

DLCI   data link channel identifier

DMIF   delivery multimedia integration framework

DNS   domain name service

DPCCH   dedicated physical control channel

DPCM   differential pulse code modulation

DPDCH   dedicated physical data channel

DRQ   delete request

DRX   discontinuous reception

DS   differentiated services

DSCH   downlink shared (transport) channel

DSCP   differentiated services code-point

DTAP   direct transfer application part

DTCH   dedicated traffic channel

DTM   dual transfer mode

DTX   discontinuous transmission

EACH   enhanced access channel

ECAM   extended channel assignment message

ECSD   enhanced circuit switch data

EDGE   enhanced data rates for GSM/global evolution

EF   expedited forwarding

EFR   enhanced full rate

EIR   equipment identification register

ERL   echo return loss

ES   elementary stream

ESN   electronic serial number

ESP   encapsulating security payload

ETSI   European Telecommunications Standards Institute

EVRC   enhanced variable rate codec

FA   foreign agent

FAC   foreign agent challenge

FACH   forward access channel

FCH   fundamental channel

FCS   frame check sequence

FDD   frequency division duplex

FDMA   frequency division multiple access

FEC   forward error control; forwarding equivalence class

FER   frame error rate

FR   full rate

F-SYNC   forward synchronization (timing and framing to mobile station)

GERAN   GSM/EDGE Radio Access Network

GGSN   gateway GPRS node

GIF   graphics interchange format

GOB   group of blocks

GOP   group of pictures

GOV   group of video object planes

GMSK   Gaussian minimum shift keying

GSN   GPRS support node

GTP   GPRS Tunneling Protocol

HARQ   hybrid automatic repeat request

HEC   header error control

HLR   home location register

HSDPA   high-speed downlink packet access

HS-DSCH   high-speed downlink shared channel

HSS   home subscriber server

HS-SCH   high-speed shared control channel

HVXC   harmonic vector excitation coding

IAM   initial address message

IAR   IAM reject

ICMP   Internet Control Message Protocol

IE   information element

IEEE   Institute of Electrical Engineers

IETF   Internet Engineering Task Force

IKE   Internet key exchange

IMSI   international mobile subscriber identity

IP   Internet Protocol

IPHC   IP header compression

IRN   international roaming number

IS   interim standard

ISDN   Integrated Services Digital Network

ITU   International Telecommunications Union

IWF   interworking function

JPEG   Joint Photographic Experts Group

LD-CELP   low delay code excited linear prediction

LDP   Label Distribution Protocol

LI   length indicator

LLC logical link control

LPDP local policy decision point

LSP label-switched path

LSR label switch router

LTU logical transmission unit

MAC media access control

MC multipoint controller; motion compensated

MCC mobile country code

MCSB message control status block

MCU multipoint control unit

MGW media gateway

MGC media gateway controller

MGCF media gateway control function

MGCP Media Gateway Control Protocol

MIB management information block

MIME multipurpose Internet mail extensions

MIN mobile identification number

MM mobility management

MOS mean opinion score

MP multipoint processor

MP3   MPEG audio level 3

MPEG   Motion Pictures Experts Group

MPLS   multiprotocol label switching

MP-MLQ   multipulse, multilevel quantization

MPPC   Microsoft point-to-point compression

MRF   multimedia resource functions

MS   mobile station

MSC   mobile switching center

MSID   mobile station identifier

MSS   maximum segment size

MTP-3b   message transfer part 3b (of SS7)

MTU   maximum transmission unit

NAK   negative acknowledgment

NAS   nonaccess stratum

NBAP   Node B application part

NMT   Nordic Mobile Telephone

NNI   network-to-network interface

NSAPI   network service access point identifier

NS-VC   network service virtual connection

NS-VL   network service virtual link

OCSF   orthogonal constant spreading factor

OSI   open systems interconnect

OVSF   orthogonal variable spreading factor

PACCH   packet associated control channel

PAGCH   packet access grant channel

PAP   Password Authentication Protocol

PBCCH   packet broadcast control channel

PCCH   packet control channel

PCF   policy control function; packet control function

PCM   pulse code modulation

PCR   peak cell rate

PCU   packet control unit

PDCH   packet dedicated channel

PDCP   Packet Data Convergence Protocol

PDP   policy decision point; Packet Data Protocol

PDTCH   packet data traffic channel

PDU   protocol data unit

PEP   policy enforcement point

PFC   packet flow context

PFI   packet flow identifier

PHB   per hop behavior

PLMN   public land mobile network

PN   pseudonoise

PNCH   packet notification channel

POTS   plain old telephone service

PPCH   packet paging channel

PPP   Point-to-Point Protocol

PRACH   physical random access channel

PTM   point-to-multipoint

PTP   point-to-point

PSS   packet-switch streaming services

QCIF   quarter common intermediate format

QDU   quantization distortion unit

QoS   quality of service

RAB   radio access bearer

RACH   random access (transport) channel

RADIUS   Remote Authentication Dial-In User Service

RANAP   radio access network application part

RAS   registration, admission, and status

RC   radio configuration

RCPC   rate compatible punctured convolutional code

RF   radio frequency

RFC   request for comment

RLC   radio link control

RLP   Radio Link Protocol

RM   resource management; resynchronization marker

RNC   radio network controller

ROHC   robust header compression

RRC   radio resource control

RSVP   Resource Reservation Protocol

RTCP   Real-Time Control Protocol

RTP   Real-Time Protocol

RTSP   Real-Time Streaming Protocol

RTT   round-trip time

RWND   receive window

SAAL   signaling ATM adaptation layer

SAPI   service access point identifier

SCAM   supplemental channel assignment message

SCAMM   supplemental channel assignment minimessage

SCCH   supplementary code channel

SCCP   signaling connection control part (of SS7)

SCH   supplementary channel; synchronization channel

SCR   sustained cell rate

SCRM   supplemental channel request message

SCTP   Stream Control Transmission Protocol

SDH   synchronous digital hierarchy

SDP   Session Description Protocol

SDU   service data unit

SGSN   serving GPRS support node

SIF   source image format

SIP   Session Initiation Protocol

SLA   service level agreement

SMpSDU   support mode for predefined SDU size

SMS   short message service

SN   sequence number

SNDCP   Subnetwork Data Convergence Protocol

SNMP   Simple Network Management Protocol

SNR   signal-to-noise ratio

SNS   sequence number space

SO   service option

SONET   Synchronized Optical Network

SQCIF   sub-QCIF

SRB   signaling radio burst

SS7   Signaling System Seven

SSC   sychronization complete; slice structure method

SSCF   service-specific coordination function

SSCOP   service-specific connection part

SSCS   service-specific convergence sublayer

SSN   starting sequence number

SSRC   synchronization source

TACS   Total Access Communications System

TB   transport block

TBF   temporary block flow

TBS   transport block set

TC   transaction capability

TCH/F   traffic channel/full-rate

TCH/H   traffic channel/half-rate

TCP   Transmission Control Protocol

TCTF   transport channel type field

TDD   time division duplex

TDMA   time division multiple access

TEID   (GTP) tunnel endpoint identifier

TF   transport format

TFCI   transport format combination indicator

TFCS   transport format combination set

TFI   temporary flow identity (GPRS); transport format indicator (WCDMA)

TFO   tandem free operation

TFRI   transport format and resource indicator

TFT   traffic flow template

TLLI   temporary logical link identity

TM   transparent mode

TMSI   temporary mobile subscriber identity

TOS   type of service

TPC   transmit power control

TrFO   transcoder free operation

TSN   transmission sequence number

TTI   transmission time interval

UDP   User Datagram Protocol

UE   user equipment

UL   uplink

UM   unacknowledged mode

UMTS   Universal Mobile Telecommunications Service

URL   uniform resource locator

USF   uplink state flag

UTRAN   UMTS Terrestrial Radio Access Network

UUI   user-to-user information

UWCC   Universal Wireless Communications Consortium

VBR   variable bit rate

VBS   voice broadcast service

VGCS   voice group-call service

VLR   visitor location register

VoFR   Voice over Frame Relay

VoIP   Voice over IP

VOP   video object plane

VRC   video redundancy coding

VTC   visual texture coding

WAP   Wireless Application Protocol

WCDMA   wideband code division multiple access

WS   window size

XHTML   extensible hypertext markup language

XOR   exclusive OR

About the Author

Robert Lloyd-Evans earned a Ph.D. from Cambridge University in 1968 and has worked in industry on data communications for 20 years. Initially, he wrote communications software and programs for the design of wide area networks; subsequently, he designed numerous corporate networks based on X25, frame relay, and TCP/IP. He has a special interest in QoS issues and has previously written a book entitled *Wide Area Network Design and Optimization* for Addison-Wesley Publishers.

Index

Recent Titles in the Artech House Mobile Communications Series

John Walker, Series Editor

Wireless Intelligent Networking, Gerry Christensen,
  Paul G. Florack, and Robert Duncan

Wireless LAN Standards and Applications, Asunción Santamaría
  and Francisco J. López-Hernández, editors

Wireless Technician's Handbook, Andrew Miceli

For further information on these and other Artech House titles,
including previously considered out-of-print books now available
through our In-Print-Forever® (IPF®) program, contact:

Artech House   Artech House
685 Canton Street   46 Gillingham Street
Norwood, MA 02062   London SW1V 1AH UK
Phone: 781-769-9750   Phone: +44 (0)20 7596-8750
Fax: 781-769-6334   Fax: +44 (0)20 7630-0166
e-mail: artech@artechhouse.com   e-mail: artech-uk@artechhouse.com

Find us on the World Wide Web at:
www.artechhouse.com

The Artech House Universal Personal Communications Series

Ramjee Prasad, Series Editor

CDMA for Wireless Personal Communications, Ramjee Prasad

IP/ATM Mobile Satellite Networks, John Farserotu and Ramjee Prasad

OFDM for Wireless Multimedia Communications, Richard van Nee and Ramjee Prasad

Radio over Fiber Technologies for Mobile Communications Networks, Hamed Al-Raweshidy and Shozo Komaki, editors

Simulation and Software Radio for Mobile Communications, Hiroshi Harada and Ramjee Prasad

Third Generation Mobile Communication Systems, Ramjee Prasad, Werner Mohr, and Walter Konhäuser, editors

Towards a Global 3G System: Advanced Mobile Communications in Europe, Volume 1, Ramjee Prasad, editor

Towards a Global 3G System: Advanced Mobile Communications in Europe, Volume 2, Ramjee Prasad, editor

Universal Wireless Personal Communications, Ramjee Prasad

WCDMA: Towards IP Mobility and Mobile Internet, Tero Ojanperä and Ramjee Prasad, editors

Wideband CDMA for Third Generation Mobile Communications, Tero Ojanperä and Ramjee Prasad, editors

WLAN Systems and Wireless IP for Next Generation Communications, Neeli Prasad and Anand Prasad, editors